Environmental Division 2018

Core Programming Area at the 2018 AIChE Spring Meeting and 14th Global Congress on Process Safety

Orlando, Florida, USA
22 – 25 April 2018

ISBN: 978-1-5108-6443-6

Printed from e-media with permission by:

Curran Associates, Inc.
57 Morehouse Lane
Red Hook, NY 12571

CURRAN ASSOCIATES INC. proceedings.com

Some format issues inherent in the e-media version may also appear in this print version.

Printed by Curran Associates, Inc. (2018)

AIChE
120 Wall Street, FL 23
New York, NY 10005-4020

Phone: (800) 242-4363
Fax: (203) 775-5177

www.aiche.org

Additional copies of this publication are available from:

Curran Associates, Inc.
57 Morehouse Lane
Red Hook, NY 12571 USA
Phone: 845-758-0400
Fax: 845-758-2633
Email: curran@proceedings.com
Web: www.proceedings.com

TABLE OF CONTENTS

512123 Flameless Oxidation for the Heating of Ethylene Furnaces

Abstract:

the Proprietary Technology Flameless Oxidation (FLOX) of the Ws-Company in Stuttgart/Germany Is Successfully Applied As Special Burners Since Almost 30 Years in Steel Treatment Furnaces with Installed Burners Well Above 50000 so Far. a Particular Advantage of FLOX Is the Extreemly Low Nox Emissions in the Range of 20 to 30 Mg/Nm3 at Very Low Noise Levels and with a Very Robust Operation Even upon Highly Fluctuating Fuelgas Compositions.a New Approach Has Been Taken to Install Such Burners in Cracking Furnaces of Ethylene Plants. Based on a Good Amount of Prework Showed That the Above Advantages Can be Applied to Cracking Furnaces As Well While Keeping the Thermal Efficiency of the Furnace Similar to Furnaces with Conventional Burners, Meaning Similar Fuelgas Consumption. Currently a Large Petrochemical Operator Is Preparing the Conversion of Several Existing Cracking Furnaces to FLOX-Burners. Successful Installation and Operation Is Expected to Take Place in the 1st Half of 2018.the Author of This Abstract Dr. Sebastian Muschelknautz (CEO & owner of Takeoff Engineering Ltd.) Is Cooperating with Ws-Company. the CEO and Owner of Ws-Company Dr. Joachim Wünnung Will be the Coauthor of This Abstract and Presentation

506757 Simultaneous Desulfurization and Denitrification of Flue Gas with Haloalkaliphilic Bacteria

Abstract:

Sulfur dioxide and nitrogen oxides from flue gas are major pollutants in air. Haloalkaliphilic microorganisms are used to treat flue gas in this study.

The process of sulfate reduction by haloalkaliphilic microorganisms was feasible. Compared with lactate, ethanol was a more appropriate electron donor for haloalkaliphilic sulfate-reducing bacteria (SRB). When ethanol was used as sole electron donor with $COD/SO_4^{2\text{â`'}}$ ratio of 4.0, the optimum HRT was 18 h with a sulfate removal efficiency of 97.8% and a sulfate removal rate of 6.26 kg/m^3 d. There was no sulfide inhibition in haloalkaliphilic sulfate-reducing bioreactor. Based on denaturing gradient gel electrophoresis analysis of 16S rRNA, the major SRB was *Desulfonatronovibrio* sp. and *Desulfonatronum* sp.

The effects of nitrate and salinity on the performance of haloalkaliphilic sulfate-reducing bioreactors were investigated. It was found that when COD/SO_4^{2-} and HRT were 2.35 and 24 h, respectively, the addition of nitrate did not affect the performance of sulfate-reducing bioreactor when nitrate concentration was below 1000 mg/L. This result indicated that the process of haloalkaliphilic simultaneous desulfurization and denitrification was feasible. When nitrate concentration reached 1500 mg/L, sulfate reduction was obviously inhibited by nitrate. Activities of other microorganisms were also inhibited by nitrate. It was found that concentration of sulfate reduced at the bottom of bioreactor accounted for 62.1% of total concentration of sulfate reduced in bioreactor.

When HRT was 24 h, the optimum $NO_3^{\text{â`'}}/SO_4^{2\text{â`'}}$ ratio was 3.0 with a nitrate removal rate of 6.0 kg/m^3 d and a sulfate removal rate of 1.39 kg/m^3 d, respectively. There was no sulfide inhibition in bioreactor. Lactate was a more appropriate electron donor for denitrifying bacteria (DB), while SRB preferred to utilize ethanol. Based on denaturing gradient gel electrophoresis analysis of 16S rRNA, the major SRB and DB were *Desulfonatronovibrio sp.* and *Halomonas campisalis*, respectively. Decrease in NO_3^-/SO_4^{2-} ratio led to obvious changes in bacterial community. Although the SRB became dominant, the population of DB also increased.

These results indicated that haloalkaliphilic microorganisms can reduce sulfate and nitrate simultaneously at high salinity with high absorption efficiency and low cost.

512745 A Case Study on Building Safety: The Coupling of Indoor and Outdoor Air Quality Modeling for Addressing Nearby Terrorist Chemical Attacks on Building Safety

Abstract:

Terrorist attacks using chemical or biological agents represent an increasing threat to building safety. Impacts from these types of attacks are of great concern to architects and engineers designing new buildings as well as administrators of existing structures.

Air dispersion modeling can be used to assess this threat. An "outdoor" air dispersion model can be used to predict the dispersion of a given agent through the atmosphere, while an "indoor" air dispersion model can be used to predict the transport of that same agent through a building given the geometry of the building and its ventilation system.

A case study illustrating modeling of a fictitious terrorist attack will be conducted, using a combination of the models AERMOD and FDS to simulate dispersion through both the atmosphere and within the building. The inputs required to conduct such an analysis and the conclusions/recommendation that can be made will be summarized.

513716 Justifying Energy Savings in Ethylene Crackers in a Low Energy Price Environment

Abstract:

Global energy prices have dropped to their lowest level in a decade and have taken with them the days of easy justification for energy improvements. In specific countries where energy prices are still subsidised, this is a familiar situation in everyday energy management. The subject of this paper is the different ways that have been used to improve the economics of energy projects in low price situations in order to make viable project investments.

Case studies will be used to illustrate such methods as:

- Debottlenecking energy critical equipment to increase throughput,
- Overcoming an energy cap to improve severity and product yield,
- Making energy savings to meet legislative targets, and
- Using energy reduction to justify changing catalysts to improve yield.

511653 Retrospective Health Impact Assessment for Pm 2.5 Pollution in Mexico City 1990-2015

Abstract:

Fine particulate matter has a well-documented association to mortality. We have conducted a retrospective health impact assessment to quantify the lives saved in Mexico City from 1990-2015 due to PM 2.5 reduction resulting from pollution control policies, utilizing the Integrated Exposure Response model developed for the Global Burden of Disease study. In addition, we monetized the health benefits using city-specific value of statistical life (VSL). Our analysis suggests that 21,000 lives saved during that period and benefits in the order of 18 billion 2015 USD. We also developed a novel model to retrospectively decompose the health benefits and attribute them to one policy utilizing sulfur reduction in fuel policy as an example. Our model indicates that this policy has resulted in 1,600 lives saved. This suggests that Mexico City air pollution policies could serve as a model for developing countries and the industry with environmental regulations at infancy stage.

512814 Sustainability Analytics - the Critical Link between Sustainability and Business Strategy for Actionable Insights

Abstract:

Advances in Enterprise Systems Are Making It Feasible for Corporations to Track and Transform Sustainability Performance. the Materiality of These Seemingly Non-Economic Impacts Is the Critical Link between Sustainability and Business Strategy. Business Analytics Enable Organizations to Convert Raw Data into Actionable Insights to Achieve Their Sustainability Goals. Sustainability Analytics Analyze the Data and Provide Insights That Help Transform Sustainability Information into Action across the Value Chain and Life Cycle. This paper will Provide Business Leaders with Insight into How to Determine Which Sustainability Metrics Are Material to Them and Relevant to Their Business, and How to Incorporate Them into Corporate Strategy, Resource Allocation Prioritization and Decision Making Process.

Based on My Book Introduction to Sustainability Analytics Under Publication By CRC Press

512839 The Reaction Mechanism and Kinetics of Magnesium Oxide for CO_2 Immobilization

Abstract:

The CO_2 immobilization is significant for the sustainable energy and chemistry industries. And the magnesium oxide is a clean and economic agent using for the CO_2 immobilization. Therefore, the reaction of CO_2 with magnesium oxide suspended in water was investigated, the mechanism proposed, and the kinetics regressed as well. The CO_2 pressures from 1 to 5 (MPa), as well as ratios of solid to liquid, stirring speeds and temperatures, significantly influenced the dissolution of magnesium oxide. And as high as 26.63 g/L MgO was dissolved in the solution under the condition of CO_2 pressure 5 MPa at 50 °C.

In the process, the magnesium oxide became magnesium hydroxide through hydration reaction, and then the hydroxide dissolved by carbonated water. In the carried out reaction of CO_2 with magnesium hydroxide suspension, no solid product was found covering the solid $Mg(OH)_2$. Therefore, a reaction mechanism based on the hydration of Mg^{2+} was proposed, where the concept of â€˜progressive liquid filmâ€™ was introduced to the reaction system. And the derived kinetic model for the dissolution was fitted the experiment data well. The pressurized magnesium hydroxide dissolution was found to be rate-controlled by the diffusion in the progressive liquid film around the solid material. The apparent activation energy, obtained from 17.1 to 25.4 (kJ/mol) in the experiments, indicated also that the dissolution inclined to be diffusion rate controlled. Moreover, the apparent diffusion coefficients of H^+ in the liquid film were calculated and an empirical expression was obtained for the first time. The kinetic model might be applicable to other dissolution reactions similarly with just soluble product.

513421 Wsa Technology - a Competitive Solution for Refinery Sulfur Management

Abstract:

For more than 30 years Haldor Topsoe has designed plants for purification of sulfur containing waste gases derived from a wide variety of process industries.

The WSA (Wet gas Sulfuric Acid) Technology for handling sulfur gases in a refinery is significantly different from those offered by other companies. While competing technologies produce elemental sulfur, the WSA process recovers sulfur from H_2S or other sulfur compounds in the form of commercial grade sulfuric acid. The end product can be used in the alkylation unit or sold to fertilizer and metallurgical industries. Furthermore, our process is flexible and can process other refinery waste streams like SWS gas, Claus tail gas and spent alkylation acid, offering an integrated solution for sulfur management.

The WSA technology is also characterized by high sulfur removal efficiency with more than 99% sulfur recovery, up to 99.99% when required, and is always in compliance with emission legislation. The process does not consume any chemicals nor will there be any solid or liquid effluents requiring additional treatment. This in combination with very low cooling water consumption makes WSA technology a very competitive solution for treatment of waste gases.

Our technology has proven feasible for treating sulfurous gases in many different applications, and today more than 150 units have been contracted worldwide handling a wide range of gas flows and with daily sulfuric acid production ranging from 4 to 1,140 MTPD.

The presentation will explain the basic principles of the WSA technology and the alternatives for refinery applications.

Research on Gas-Liquid Coalescence Filtration Performance of Electrospun Polyacrylonitrile Nanofiber Composite Filters

Feng Chen, Zhongli Ji, Qiangqiang Qi
Beijing Key Laboratory of Process Fluid Filtration and Separation, College of Mechanical and Transportation Engineering, China University of Petroleum, Beijing, PR China

Abstract

Coalescing filters are widely used to remove small droplets from gas streams, while it is increasingly difficult to meet the pollution emission requirements with traditional microfiber filters. Due to a large fiber surface area and small fiber diameter, the nanofiber materials are able to improve the overall filtration performance of filters. Electrospinning is one of the most common and available methods to prepare nanofibers. The work presents a study on the filtration performance of composite nanofiber filters with different properties, aiming to provide a better understanding of the application of nanofibers in the gas-liquid filtration field.

A range of solutions with four different concentrations were prepared, with a polymer of Polyacrylonitrile (PAN) and a solvent of N, N-dimethylformamide (DMF). The glass fiber materials were selected as a substrate, on which the nanofibers were collected. Four different voltages were applied for each solution concentration. An additional layer of glass fiber, used as a protective layer, was covered on the prepared nanofiber layer to construct a sandwich. The effects of solution concentration and electrospun voltage on the structure and morphology of nanofibers, as well as filtration performance of composite filters were investigated experimentally. A scanning mobility particle sizer (SMPS) was used to measure the aerosol concentrations upstream and downstream of the filters, from which the efficiency was derived.

The results showed that the nanofibers prepared with various concentrations possessed different morphologies, where the overlap of two or three fibers occurred for high concentrations while beads were present for low concentrations. When the concentration was 14%, the fiber diameter first decreased, then increased with increasing voltage. It was also found that the solution concentration had a more significant effect on the fiber diameter, compared with the applied voltage. A low solution concentration had a small fiber diameter, leading to a high equilibrium efficiency with an unremarkable increase in pressure drop.

Keywords: Aerosol; Filtration; Particle

1. Introduction

Filtration of liquid droplets from gas streams is of significant importance to many industrial process applications such as compressed air, natural gas, and engine crankcases [1-3]. Glass microfiber filter is one of the most common and efficient methods to remove such liquid particles, especially for those with diameter in micrometer size range. However, for sub-micrometer sized liquid particles, it increasingly reveals that conventional glass fiber filters cannot meet the requirement of high-efficiency filtration, which can possibly be achieved using nanofiber filters due to their high fiber surface area, large porosity, and small fiber diameter [4]. There are many different ways to prepare nanofibers, including electrospinning, meltblown, and sea-island, etc. [5,6]. As a range of polymers can be used to prepare the nanofibers and comparatively simple operation, electrospinning has been one of the most popular methods to fabricate nanofibers or composite nanofiber filters [7].

A number of studies have been conducted on the research and applications of polyacrylonitrile (PAN) nanofibers in the gas-solid filtration field [8,9]. Yun et al. [10] prepared a composite filter consisting of a substrate of polyester fibers and a PAN nanofiber layer, and this composite filter presented a better performance than commercial glass microfiber filters. Matulevicius et al. [11] prepared a lot of nanofiber membranes with different polymers through electrospinning, and found that PAN nanofiber membrane had a relatively higher efficiency as compared to other materials.

Furthermore, some work have also been done on the application of nanofiber in the field of gas-liquid filtration. Patel et al. [12] prepared a composite filter with glass fiber and nylon nanofiber to investigate the effect of nanofiber diameter on gas-liquid filtration performance of the composite filter. Hajra et al. [13] prepared a composite filter with electrospun polyamide nanofibers on the substrate mat, with the results indicating that the filtration performance of filter would increase with increasing amount of nanofibers in composite filters.

However, few studies in the literature have reported the effect of preparing parameters of nanofibers on the filtration performance of gas-liquid coalescing filters. Therefore, in this paper, a series of composite filters were prepared, with different polymer solution concentrations and voltages used during the electrospinning process. The fabricated filters were characterized using the scanning electron microscope. Finally, the effects of polymer solution concentration and electrospun voltage on gas-liquid filtration performance of composite filters were analyzed.

2. Experimental

2.1. Materials and filter preparation

Polyacrylonitrile powders with molecular weight of 150,000 were purchased from Sigma Aldrich. N, N-dimethylformamide (DMF, AR grade) was purchased from Allladin (Shanghai, China). These materials were used without further purification.

Fig. 1 shows the schematic diagram of electrospinning apparatus. Different solutions for electrospinning were prepared by adding various amount of PAN powders into the solvent of DMF with concentrations controlled at 8%, 10%, 12%, and 14% (w/v), respectively. The solution was put into a 5 mL syringe after stirring for more than 12 h with rotation speed set at 1000 r/min at room temperature. A metal needle (22G, 0.4 mm inner diameter) in front of syringe was used to connect with the positive high voltage source. Four different voltages of 8, 12, 16, and 20 kV, were selected for each concentration of solutions to produce the nanofibers with different diameters. A collector located in front of the needle, with a fixed distance of 20 cm to the needle tip, was rotating in a speed of 40 r/min during the electrospinning process. Various pumping velocities for the solution to be sprayed were applied to ensure a stable jet-flow. All composite filters have an identical gram per area of the nanofiber layer, which was obtained by precisely controlling the spinning time. Note that a glass fiber layer, as a substrate, was first surrounded on the surface of collector before electrospinning, on which the nanofiber layer was randomly deposited, after that a same glass fiber layer covered on the nanofiber mat as a protective layer. Hence, a sandwich-structured filter was obtained.

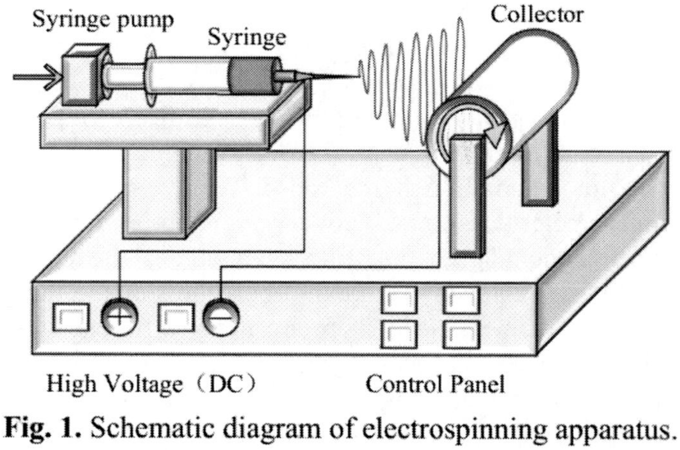

Fig. 1. Schematic diagram of electrospinning apparatus.

The naming of the prepared composite filters was based on the preparation parameters. For example, filter C8V8 represents that the nanofiber layer in such filter was prepared on the condition that the solution concentration was 8% (w/v) and applied voltage was 8 kV, similarly for the naming of other filters. Also, filter C8, C10, C12, and C14 indicate all composite filters with nanofiber layers prepared with polymer solution concentration of 8%, 10%, 12%, and 14% (w/v), respectively.

2.2. Filtration experimental setup

Fig. 2 shows the schematic diagram of filtration experimental setup. As shown, the liquid aerosol was generated through an aerosol generator (9306A, TSI). A mixing chamber located after the generator was used to mix the clean air which was purified by a HEPA (high efficiency particulate air) filter and the generated aerosol. Then the mixed aerosol horizontally passed through the test filter, which was vertically assembled in the filter chamber. A differential pressure transmitter (EJX-110A, Yokogawa) was used to monitor and record the pressure drop throughout the experiment. A scanning mobility particle sizer (Model 3936, TSI) was used to measure the aerosol concentrations upstream and downstream of the test filter. Finally, the filtered gas was pumped through a vacuum pump and the flow rate throughout each experiment was controlled at 0.12 m/s by a mass flow controller placed front of the pump. The parameters of measuring equipment used are listed in Table 1.

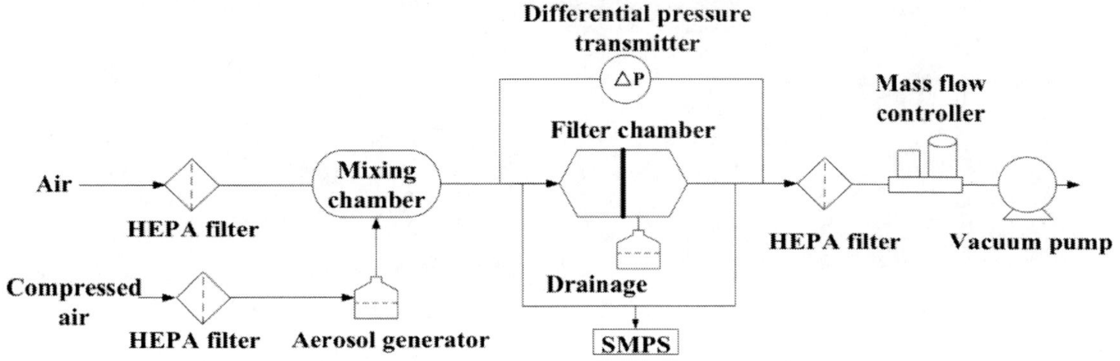

Fig. 2. Schematic diagram of filtration experimental setup.

Table 1. Parameters of measuring equipment

Device	Model	Accuracy	Range
Aerosol generator	9306 A	±2%	0–5 μm
Differential pressure transmitter	EJX-110A	±0.04%	0–20 kPa
SMPS	3936	±4%	0.05–0.8 μm
Mass flow controller	MCR 500 slpm	±0.4%	0–500 slpm

Filtration efficiency of different filters was derived from the upstream and downstream aerosol concentrations and defined by

$$\eta = 1 - Pe = 1 - \frac{C_{\text{out}}}{C_{\text{in}}} \tag{1}$$

where η is the filtration efficiency of the filter, Pe is the penetration of the filter, and C_{out} and C_{in} are the downstream and upstream aerosol concentrations of the filter, respectively.

3. Results and discussion

Fig. 3 shows the scanning electron microscope (SEM) images of four typical nanofiber layers prepared with different polymer solution concentrations. As shown, the obtained fiber structures were highly dependent on the polymer solution concentration. When the concentration was low at 8% (w/v), it appears that some spindle-shaped fibers or beads were present, while the fibers became more uniform when polymer solution concentration increased to 10% and 12% (w/v). Surprisingly, an overlap of two or three fibers appeared after the polymer solution concentration increased to 14% (w/v). Fig. 3 also shows that the fiber diameter increased significantly with increasing polymer solution concentration.

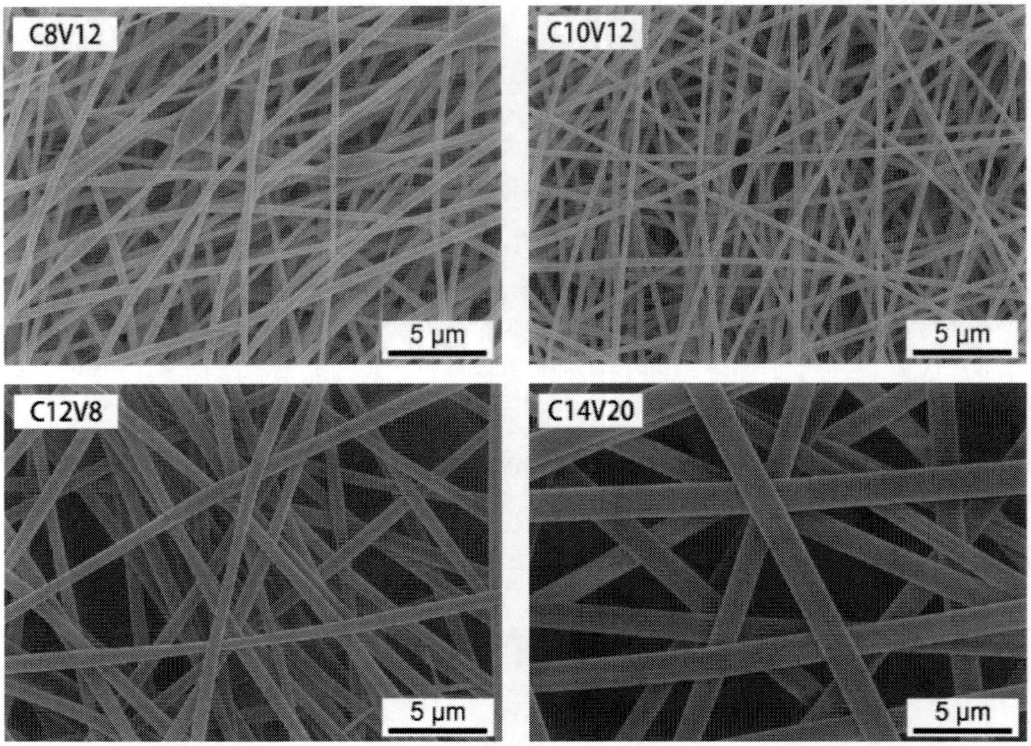

Fig. 3. SEM images of nanofibers under different solution concentrations.

Fig. 4 shows the average fiber diameters of different nanofiber layers. The results were obtained through statistically analyzing the corresponding SEM images. As shown, other than filter C14, the voltage had little effect on the fiber diameter. For the filter C14, it seems that the fiber diameter first decreased with increasing voltage then slowly increased with continually growing voltages.

Overall, it can be recognized that the solution concentration had a more pronounced impact on the fiber diameter of nanofibers than applied voltage. For instance, when polymer solution concentration was 14% (w/v), the fiber diameter obtained was approximately 1500 μm, about 6 times that of nanofibers prepared with polymer solution concentration of 8% (w/v).

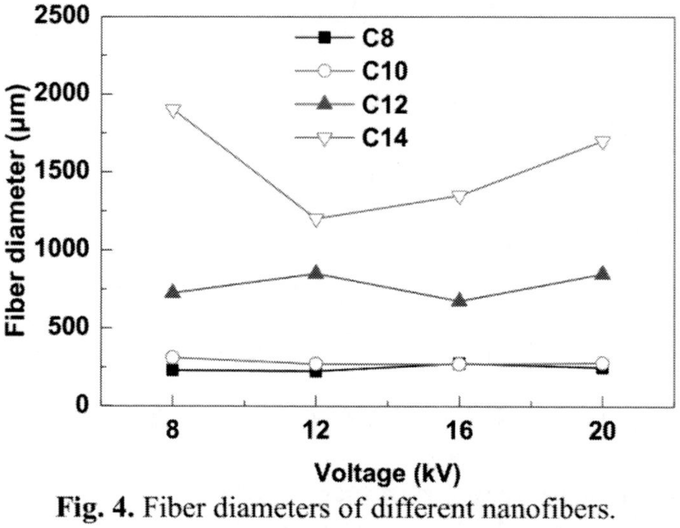

Fig. 4. Fiber diameters of different nanofibers.

Fig. 5 shows the typical pressure drop and penetration evolutions of the two-layer substrate glass fiber filter, which was regarded as the control filter in this work. As shown, the filtration process can be defined into three stages according to the pressure drop profile. Stage 1, 2, and 3 corresponded to the channel stage, jump stage, and steady-state stage, respectively, which was first proposed by Kampa et al. [14]. Because most filters used in industrial applications operated in the steady state (Stage 3), following sections will focus on the comparison of filtration performance of filters after reaching the steady state.

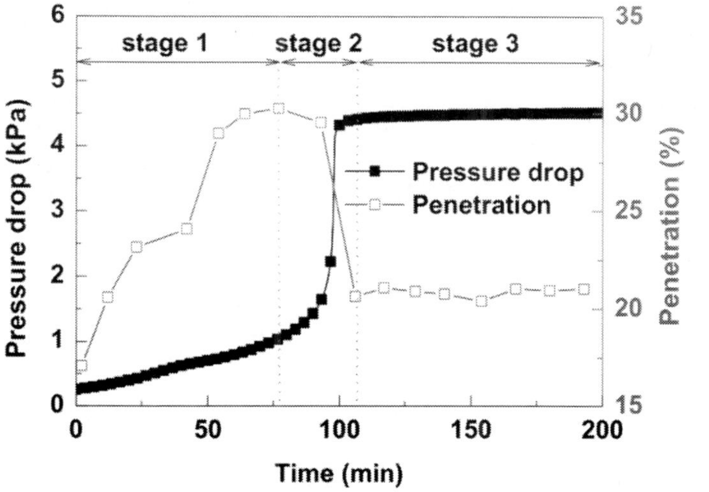

Fig. 5. Pressure drop and penetration evolutions of the control filter.

Fig. 6 shows the steady-state pressure drop of different composite filters. As shown, there existed no significant growth in the pressure drop of composite filters when compared to the control filter. Also, it is easy to be found that the applied voltage had little effect on the steady-state pressure drop of filters. Besides, filter C10 had a slightly higher pressure drop when in 16 and 20 kV, which was not fully understood, but likely due to the comparatively poor reproducibility of preparation process of nanofibers.

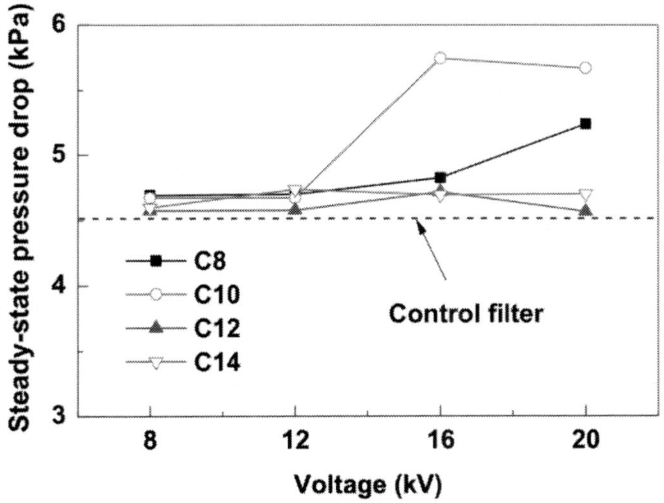

Fig. 6. Steady-state pressure drop of different composite filters.

Fig. 7 shows the steady-state filtration efficiency of different composite filters. It is obvious that the filtration efficiency was largely improved after adding nanofibers between two glass fiber layers. Filter C10 had almost the highest efficiencies at each voltage, while the filter with highest polymer solution concentration (filter C14) presented the lowest efficiencies at each voltage. Therefore, it can be concluded that the filter constructed with nanofibers that were prepared with smaller polymer solution concentration exhibited a higher filtration efficiency, due to smaller fiber diameter as shown in Fig. 4. Based on the little change in the steady-state pressure drop of composite filters as compared to the control filter, it is summarized that the composite filters prepared under lower polymer solution concentration had a better overall filtration performance.

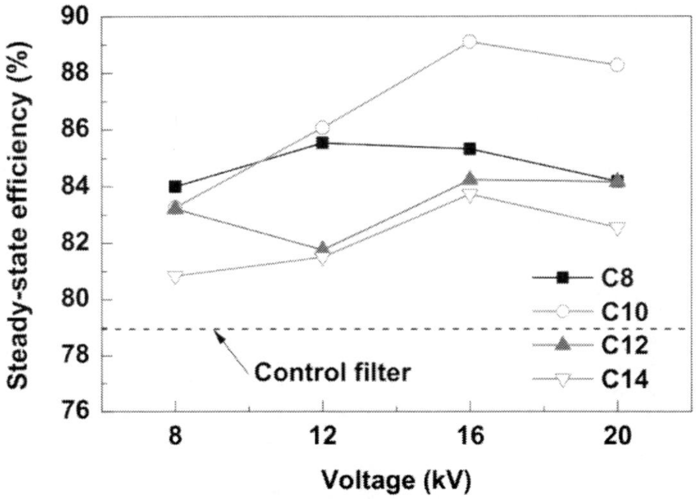

Fig. 7. Steady-state filtration efficiency of different composite filters.

4. Conclusions

In this work, the effects of polymer solution concentration and applied voltage during electrospinning process on the filtration performance of composite filters were experimentally studied. The results showed that the polymer solution concentration had a greater influence on the nanofiber diameter and steady-state filtration efficiency of composite filters than applied voltage, while both of these two parameters had little effect on steady-state pressure drop. Filters composed of nanofibers that were prepared with lower polymer solution concentration showed a better overall filtration performance. In summary, the composite nanofiber filter was a potentially good choice to be used in the field of gas-liquid filtration.

Acknowledgments

This material is based upon work supported by National Natural Science Foundation of China under grant No. 51376196.

References

1. Mead-Hunter R., King A.J.C., Mullins B.J., Aerosol-mist coalescing filters – A review, Separation and Purification Technology, 2014, 133: 484-506.
2. Mullins B.J., Mead-Hunter R., Pitta R.N., *et al.*, Comparative performance of philic and phobic oil-mist filters, AIChE Journal, 2014, 60(8): 2976-2984.
3. Chang C., Ji Z.L., Zeng F.Y., The effect of a drainage layer on filtration performance of coalescing filters, Separation and Purification Technology, 2016, 170: 370-376.
4. Zhang Q., Welch J., Park H., *et al.*, Improvement in nanofiber filtration by multiple thin layers of nanofiber mats, Journal of Aerosol Science, 2010, 41(2): 230-236.
5. Chen R.Z., Hu Y., Yuan J.G., *et al.*, Preparation of electrospinning MnO2/PAN nanofibers and catalytic oxidation on formaldehyde, Journal of Textile Research, 2015, 36(5): 1-6.
6. Ma X.L., Zhang L.Y., Li H.Y., *et al.*, Preparation of oriented nanowires by melt differential electrospinning, Journal of Textile Research, 2017, 38(1): 8-12.
7. Matulevicius J., Kliucininkas L., Prasau-Skas T., *et al.*, The comparative study of aerosol filtration by electrospun polyamide, polyvinyl acetate, polyacrylonitrile and cellulose acetate nanofiber

media, Journal of Aerosol Science, 2016, 92(2): 27-37.
8. Balgis R., Kartikowati C.W., Ogi T., *et al.*, Synthesis and evaluation of straight and bead-free nanofibers for improved aerosol filtration, Chemical Engineering Science, 2015, 137: 947-954.
9. Shen L.D., Yu X.F., Cheng C., *et al.*, High filtration performance thin nanofibrous composite membrane pre-pared by electrospraying technique and hot-pressing treatment, Journal of Membrane Science, 2016, 499: 470-479.
10. Yun K.M., Hogan C.J., Matsubayashi Y., *et al.*, Nanoparticle filtration by electrospun polymer fibers, Chemical Engineering Science, 2007, 62(17): 4751-4759.
11. Matulevicius J., Kliucininkas L., Prasau-Skas T., *et al.*, The comparative study of aerosol filtration by electrospun polyamide, polyvinyl acetate, polyacrylonitrile and cellulose acetate nanofiber media, Journal of Aerosol Science, 2016, 92(2): 27-37.
12. Patel S.U., Kulkarni P.S., Patel S.U., *et al.*, Glass fiber coalescing filter media augmented with polymeric submicron fibers and modified with angled drainage channels, Separation and Purification Technology, 2013, 120: 230-238.
13. Hajra M.G., Mehta K., Chase G.G., Effects of humidity, temperature, and nanofibers on drop coalescence in glass fiber media, Separation and Purification Technology, 2003, 30: 79-88.
14. Kampa D., Wurster S., Buzengeiger J., *et al.*, Pressure drop and liquid transport through coalescence filter media used for oil mist filtration, International Journal of Multiphase Flow, 2014, 58: 313-324.

513640 Dynamic Simulation and Optimization of Turnaround Operations Among Multiple Olefin Plants for Flare Minimization

Abstract:

Flare minimization (FM) in large-scale industry is a win-win practice, which benefits not only environmental sustainability but also industrial profitability. By far, available studies have almost exclusively focused on FM problems for a single plant. However, there are few systematic FM studies focusing on FM solutions among multiple chemical plants. Material exchanges (ME) among chemical plants provide an augmented freedom for material savings, energy saving and flare source reductions. For example, plant A can borrow materials from plant B with normal operating conditions to support the start-up of plant A; meanwhile, pant B can adopt off-spec products from plant A undergoing the start-up operations to recover resources for producing more potential products. Such ME operations could be designed, scheduled, and optimized through large-scale dynamic simulation studies.

In this study, dynamic simulation and optimization have been performed for FM via ME between two olefin plants.

For the ordinary start-up of an olefin plant, no any external flows will be imported, a large amount of flaring will be generated during cracked gas compressor (CGC) start-up, because the charge gas from fired-up furnaces to run CGC needs some time to reach the required flow rate. Another flare source is from the reactor, where the outlet of the reactor should be on spec before sent to the downstream. In this work, ME among three plants are studied. Plant A is at normal-working status, plants B and C are both preparing to start-up. Plants B and C could import certain flows from plant A to reduce the number of furnaces fired-up before their CGC start-ups. Meanwhile, plants B and C also export off-spec products to Plant A to avoid flaring. The amount of these material to exchanges could vary, and the start-up schedules of two plants could be different. Therefore, a series of dynamic simulations have been done to find the optimal design of start-up schedule and ME amounts, which not only generate least flaring, but also ensure plant Aâ€™s normal production and smooth starting-up of plants B and C. The obtained strategies and simulation results will provide an in-depth understanding on synergies of three plants ME operations for flare minimizations.

Pollution, Control Status and Strategy of Volatile Organic Compounds in China's Petrochemical and Refining Industries

Jia Ke, China University of Petroleum (East China), Qingdao, P R China
Dongfeng Zhao, China University of Petroleum (East China), Qingdao, P R China; Center for Safety, Environmental, and Energy Conservation Technology, Qingdao, P R China
Yi Liu, China University of Petroleum (East China), Qingdao, P R China; Center for Safety, Environmental, and Energy Conservation Technology, Qingdao, P R China

Abstract

With rapid industrialization and urbanization in China, haze and photochemical smog, caused by fine particulate matters ($PM_{2.5}$, PM_{10}) and ozone (O_3), have become one of the most severe environmental pollution issues in China, which were seriously affecting the daily life and physical health of residents. Thus, as an important precursor of fine particulate matters and ozone, volatile organic compounds (VOCs) are now included in the national control of key pollutants and total control factors. Because of its strategic importance in industrial system and its huge VOCs emissions (accounts for 34.3% of the total emissions), petrochemical and refining industries have recently elicited public concern and given new impetus to scientific interest.

This paper reviews the VOCs emissions, and summarizes the policies, regulations, emission standards and the existing pollutant permit system for controlling VOCs in China's petrochemical and refining industries. This paper also analysis the main problems in current petrochemical and refining industries VOCs control status and management. Some countermeasures and suggestions, such as developing and improving VOCs monitoring methods, establishing VOCs emission list database, researching the best feasible control technology in petrochemical and refining industries and so on, are proposed to strengthen the control and management of VOCs emissions. Besides, the prospects and recommendations are put forward for China's future open, whole-process VOCs management model. The findings not only can provide a scientific basis for enhancing and implementation of existing VOC control strategies, but also can provide technical support for governments and enterprises to improve environmental quality and alleviate regional air pollution.

Keywords: Petrochemical and Refining Industries; Volatile Organic Compounds; Policies; Environmental Management

Introduction

With the rapid development of China's industrialization process, air pollution type has shifted from traditional coal smoke pollution, which dominated by dust and SO_2 emissions, to a composited air pollution, which dominated by fine particulate matters ($PM_{2.5}$, PM_{10}) and ozone (O_3) pollution. Air pollution is increasingly intensifying. The continuous phenomena of acid rain, haze and photochemical smog have become one of the most serious environmental issues in many cities and regions in China, seriously affecting people' daily life and physical health.

Many articles show that volatile organic compounds (VOCs) are important precursors of atmospheric ozone (O_3) and secondary organic aerosol (SOA). SOA is considered to be one of main contribution sources to $PM_{2.5}$ in the ambient air, which has contribution to the formation of the composited air pollution. Among various VOCs anthropogenic emission sources, industrial sources are the largest source of VOCs emissions. The petrochemical and refining industries, which has significant VOCs emissions among industrial sources, is undoubtedly placed on the front line of VOCs control. Therefore, VOCs research and effective control in the petrochemical and refining industries has become a hot topic in the field of atmospheric environmental management and control in China.

The status of VOC emissions in the petrochemical and refining industries, including policies, regulations, standards of VOCs control and management, were analyzed and studied. Some main problems in VOCs control current status and control management were proposed. In addition, solutions and recommendations were provided. The results can will provide technical supports to improve environmental quality.

1 China Air Quality Status

Nowadays, China has established a comprehensive national environmental air quality monitoring network and has set up more than 5,000 monitoring stations, which included 1,436 air quality monitoring stations in 338 cities at or above prefecture level (contained 135 clean control stations), 96 regional (rural) ambient air quality monitoring stations, 16 background ambient air quality monitoring stations, and 38 atmospheric particulate components/photochemical monitoring stations in 2+26 cities (i.e. Beijing, Tianjin, and 26 cities in the Beijing-Tianjin-Hebei and surrounding area).

According to the Ambient Air Quality Standards (GB 3095-2012) and the Technical Regulation for Ambient Air Quality Assessment (HJ 663-2013), six ambient air quality pollutants, i.e. SO_2, NO_2, $PM_{2.5}$, PM_{10}, O_3 and CO, were evaluated based on the ambient air quality monitoring data. The ratio of different air quality levels was assessed in different cities from 2013 to 2016. And its results were showed in Table 1. The average annual concentrations of ambient air quality pollutants from 2013 to 2016 were shown in Table 2.

Table 1 Air quality conditions from 2013 to 2016

Year	percentage of air quality standards				
	excellent and good	**slightly polluted**	**intermediately polluted**	**heavily polluted**	**very heavily polluted**
2013	60.5	22.9	8	6.2	2.4
2014	66.3	21.6	6.5	4.4	1.2
2015	76.7	15.9	4.2	2.5	0.7
2016	78.8	14.8	3.7	2	0.6

Note: Days with AQI = 0~100 indicates perfect, AQI =101~150 indicates slight pollution, AQI =151~200 indicates intermediate pollution, AQI =201~300 indicates heavy pollution and AQI >300 very serious pollution.

Table 2 Average annual concentrations of ambient air quality pollutants from 2013 to 2016

Year	**$PM_{2.5}$**	**PM_{10}**	**90th percentile concentration of O_3 daily maximum 8-hour average**	**SO_2**	**NO_2**	**95th percentile concentration of daily CO average (mg/m^3)**
2013	72	118	139	40	44	2.5
2014	62	105	140	35	38	2.2

2015	50	87	134	25	30	2.1
2016	47	82	138	22	30	1.9
Standard valve	35	70	160	60	40	4

The percentage of cities met national air quality standard was increased from 4.1% in 2013 to 24.9% in 2016, and the percentage of cities failed to meet national air quality standard was decreased from 95.9% in 2013 to 75.1% in 2016. By comparing the ambient air qualities data from 2013 to 2016, it can be found that the urban air quality has improved as a whole. It indicated that China's air pollution prevention and control has made significant progress under control and management. The annual average concentrations of $PM_{2.5}$, PM_{10}, NO_2, SO_2, and CO decreased year by year. The number of heavy pollution days in most cities has decreased. However, pollution issues, such as high-pollution loads of $PM_{2.5}$ and PM_{10} in key areas, composited pollution and heavy air pollution days which the primary pollutants were $PM_{2.5}$, O_3 and PM_{10} during these days, are still serious. The overall situation of air pollution control is still severe.

As an important precursor of SOA and ozone, VOCs have become the fourth most important atmospheric pollutants in China after sulfur dioxide, nitrogen oxides, and dust, and have been listed in national key prevention and control pollutants and total control factors. Therefore, the control and reduction of VOCs has also become one of the essential issues in the prevention and control of atmospheric pollution. In addition, petrochemical industry has attracted much attention due to its huge VOCs emission, and has become one of the key industries for VOCs management and control.

2. Volatile Organic Compounds

Volatile Organic Compounds (VOCs) are composed of hundreds of organic compounds that are directly emitted into the atmosphere. They are a class of organic compounds. For different industries, the chemical composition varies, but mainly includes organic compounds such as alkanes, unsaturated hydrocarbons, halogenated hydrocarbons, benzene series, organic acids, alcohols, and esters. For volatile organic compounds, the United States, the European Union, Japan, and other countries have been research and control VOCs decades ago. However, China starts relatively late to research and control VOCs, and is still at exploratory stage.

2.1 Definition of VOCs

There is no formal and standard definition of VOCs in the world nowadays. In china, many standards also defined VOCs, but it existed difference. The specific definitions were described in Table 3 below.

Table 3 Definition of VOCs in different standards

No.	Standard	Definition
1	Emission standard of pollutants for synthetic leather and artificial leather industry GB 21902-2008	Any organic compound (excluding methane) that has a boiling point of less than 250°C or can be discharged into the air as a gaseous molecule at atmospheric pressure
2	Emission standards of air pollutants for petroleum refining and petrochemicals manufacturing industry DB 11/447-2007	Organic compounds with a vapor pressure higher than or equal to 0.01 kPa at 20°C
3	The discharge standard of pollutants	Volatile organic compounds that saturated vapor

	for bio-pharmaceutical industry DB 31/373-2010	pressure at 0.1mmHg or above at 20°C, or melting point below room temperature and boiling point below 260°C, excluding methane
4	Indoor air quality standard GB/T 18883-2002	Volatile organic compounds that retention time between hexane and n-hexadecane analyzed by non-polar columns (polarity less than 10), which sampled by Tenax GC or Tenax TA
5	Integrated emission standard of air pollutants GB 3095-2012	Organic compounds with a vapor pressure higher than or equal to 0.01 kPa at 20°C or has corresponding volatility under specified conditions
6	Emission standard of pollutants for petroleum refining industry GB 31570-2015	Organic compounds involved in atmospheric photochemical reactions, or determined or calculated by specific methods
7	Emission standard of pollutants for petroleum chemistry industry GB 31571-2015	Organic compounds involved in atmospheric photochemical reactions, or determined or calculated by specific methods

As can be seen from Table 1, there are different definitions of VOCs in different standards. These definitions were defined in terms of volatility (such as vapor pressure, boiling point, etc.). Under this definition, VOCs basically cover all organic compounds. However, it is not clear that which substances have O_3 and $PM_{2.5}$ formation potential. They have limitations. In addition, there are some definitions that are defined based on monitoring methods. Under this definition, some of the higher melting VOCs are excluded from the definition, which cannot fully reflect the definition.

For petrochemical and refining industries, the definition is mainly based on photochemical reactivity. Under this definition, the factors can be found to control VOCs, so as to achieve the control of the formation of ozone in ambient air. It is also possible to find out specific substances with high photochemical activity at the source stage. The formation of ozone and $PM_{2.5}$ in the ambient air can be achieved by not using or instead of using such substances, or developing new processing technologies.

2.2 Sources of VOCs

The sources of VOCs include natural sources and human sources. Natural sources mainly come from plant emissions, such as plant releases, volcanic eruptions, and forest fires. They are uncontrolled. The most important emissions sources are forests and shrubs. The most important natural emissions are isoprene and monoterpenes. The human sources mainly come from the use of automobile exhaust, fuel evaporation, petrochemicals, and paints. It includes various industrial sources, agricultural sources, transportation sources, and life sources.

In China, human sources mainly come from industrial sources, transportation sources (motor vehicle emissions) and biomass combustion. Due to the rapid industrialization process, VOCs emissions from industrial sources increased year by year and accounted for more than 50% of the total VOCs emissions. Industrial emission sources are complex, involving production, usage, storage, and transportation processes. The petrochemical and refining, coatings, inks, pesticides, and automotive industries accounted for more than 80% of total industrial VOCs emissions. Research showed the VOCs emissions from petrochemical and refining industry represented about 34.3% of the total VOCs emissions.

2.3 Photochemical Reaction of VOCs

Under UV light, VOCs have a series of photochemical interactions with nitrogen oxides (NOx), which formed O_3, peroxyacetyl nitrate (PAN), highly reactive free radicals (ON, RO_2, HO_2), aldehydes and so on. These pollutants can further form secondary aerosols (SOA) through oxidation, nucleation, and condensation processes, which promotes the formation of $PM_{2.5}$.

When there is no VOCs in the air, the interactions between NO_2, NO, and O_3 in the atmosphere can reach equilibrium:

$$NO_2 \rightarrow NO + O$$

$$O + O_2 \rightarrow O_3$$

$$O_3 + NO \rightarrow NO_2 + O_2$$

When there is no VOCs in the air, the following reactions occur:

$$VOCs + OH \rightarrow HO_2$$

$$HO_2 + NO \rightarrow NO_2 + OH \text{ (reaction quickly)}$$

NO, produced by NO_2, is consumed by HO_2, resulting in NO being unable to react with ozone and breaking the equilibrium of NO_2, NO, and O_3 interactions. It eventually leading to the accumulation of O_3.

2.4 Damages of VOCs

Due to the diversity of VOCs' physical and chemical properties, the scope and impact of VOCs are also relatively widespread. VOCs are not only harmful to environmental, but also harmful to human health.

1) Impact on the environment

Some of the substances of VOCs have strong photochemical reactivity, can cause photochemical smog, and can indirectly affect the radiation balance on the Earth's surface, which in turn affects the climate. Some VOCs can destroy stratospheric ozone characteristics and reduce stratospheric ozone concentrations, such as chlorofluorocarbons (CFCs). With the diffusion and migration of VOCs pollution, the environmental problems such as acid rain, ozone layer destruction and atmospheric warming are no longer belonged to one city or country, but spread to a global scale.

2) Impact on human health

Most VOCs are toxicologically and harmful to human health. At present, more than 50 pollutants are contained in priority control pollutants list, which announced by the US EPA. VOCs can cause maladjustment of multiple systems, such as the respiratory system, nervous system and cardiovascular system. VOCs also have carcinogenicity and genetic toxicity. When the concentration of VOCs in the environment reaches a certain concentration, people may feel headaches, nausea, and vomiting. In severe cases, convulsions, coma, and even life-threatening conditions may occur.

3. Status of VOCs Emissions in Petrochemical and Refining Industries

As a pillar industry for the economy development of China, petrochemical and refining industries arises a series of environmental pollution problems, especially VOCs pollution problems during the acceleration of industry economic growth. Researchers used different estimation methods to estimate the VOCs emissions. The VOCs emissions from human-made source in China are about 20 Mt/a, and the data rapidly rising. Among the human-made source, industrial sources accounted for about 55%, in which petroleum refining and storage, organic chemicals, and fine chemicals were key industries for VOC emissions.

In November 2015, the Ministry of Ecology Environment (MEE) issued the "Guidelines for

VOCs Pollution Source Inspection Work for petrochemical and refining industries" in accordance with the theoretical basis for the classification and analysis of pollution sources in the process chain, and divided VOCs sources into 12 categories. These are the leakage of equipment static and dynamic sealing points; the volatilization loss of organic liquids storage and blending process; the volatilization loss of organic liquids loading and unloading process; the volatilization loss of wastewater in collection, storage, treatment and disposal process; organized emissions; the release from circulating cooling water system; abnormal conditions (including startup, shutdown and maintenance) emissions; flares emissions; combustion emissions; unorganized emissions; sampling process emissions and accident emissions. The document requires that all petrochemical and refining industries should basically complete the VOCs comprehensive control and establish VOCs monitoring system by 2017. The total VOCs emissions will be reduced by more than 30% from 2014. By the end of 2017, the petrochemical and refining industries have comprehensively carried out comprehensive VOCs control work.

The VOC emissions of each source from petrochemical and refining industries with different capacities are shown in Table 4.

Table 4 VOC emissions from different petrochemical and refining industries (t/a)

Source	**Sinopec Qingdao Refining Corp., Ltd.**	**Haerbin Petrochemical**	**Guangxi Petrochemical**	**Jiujiang Petrochemical**
Capacity (million t/a)	1000	420	1000	800
Leakage of equipment static and dynamic sealing points	90.2	212.9	126.3	199.4
Volatilization loss of organic liquids storage and blending process	319.1	997.3	2087.8	2027.8
Volatilization loss of organic liquids loading and unloading process	26.1	511.9	1613.9	1084.4
Volatilization loss of wastewater in collection, storage, treatment and disposal process	228.4	130.5	159.4	1061.5
Release from circulating cooling water system	225.6	124.3	121.4	853.2
Combustion emissions	7.5	19.4	221.3	13.8
Organized emissions	48.6	2.6	370.2	87.97
Unorganized emissions	472.5	0.0	0.0	167.5
Sampling process emissions	0.0	0.0	19.9	0
Flares emissions	7.2	143.6	81.7	0.02
Abnormal conditions emissions	0.0	0.0	3.0	0.4
Accident emissions	0.0	0.0	0.0	0
Total	1425.1	2142.4	4805.0	5495.9

4 VOCs Control Status

Facing the rising industrialization and deteriorating ambient air quality, China has begun to pay attention to VOCs pollution, and has gradually established a management idea, taking whole process and refinement management and control as center. VOCs management and control measures will be conducted from the source, process, and ends aspects. In an attempt to reduce air pollution and control VOCs pollution, the Chinese government has made the decision to enforce stricter regulations, including laws, standards, regulation, action plans, VOCs pollution prevention technologies and others.

4.1 Policy System

For air pollution, the two key laws are Environmental Protection Law (EPL) and Law of the People's Republic of China on the Prevention and Control of Atmospheric Pollution. As the air pollution has been shifting from typical coal smoke pollution to composited air pollution, the two laws are amended on April 2104, August 2015 respectively and implemented the next year. The current version of Prevention and Control of Atmospheric Pollution Law has expanded into the scope of legal supervision, and defined legal responsibilities of VOCs prevention and control.

Since 2010, in order to govern VOCs and improve environmental air quality, the State has successively issued various regulations and action plans, such as 'Guiding Opinions on Promoting Air Pollution Joint Prevention and Control Work to Improve Regional Air Quality', '12th Five-year Plan' on Air Pollution Prevention and Control in Key Regions', 'the State Council Air Pollution Prevention and Control Action Plan', 'Control Pollutant Release Permit System Implementation Plan'. These regulations and action plans listed VOCs as key pollutants for key air pollution prevention and control, together with particulate matter, SO_2, and NO_2. They are also placed $PM_{2.5}$ and O_3 and its precursors in a core position in the prevention and control of air pollution in China. It is proposed that by 2020, the state will establish and improve the VOCs pollution prevention and management system, and implement VOCs pollution reduction in key areas and key industries.

As one key industry, petrochemical and refining industries, the MEE issued the 'Comprehensive Plan for the Control of Volatile Organic Compounds in the Petrochemical and Refining Industries', the 'Guidelines for VOCs Pollution Source Investigation in the Petrochemical and Refining Industries' and the 'Guidelines for Petrochemical and Refining Industries Leak Detection and Restoration Work' and other documents to comprehensively control VOCs in the petrochemical and refining industries, and to greatly reduce VOCs emissions from the petrochemical and refining industries.

4.2 Emission Standards

To strengthen the control of atmospheric pollution, the National Ambient Air Quality Standard (NAAQS) (GB 3095-2012), which newly limits the ambient levels of $PM_{2.5}$ and ozone and sets clear reduction goals for air quality planners and regulators, was revised on 29 February 2012 and enacted on 1 January 2016. As for petrochemical and refining industries, some industrial emission standards have issued, like 'Emission Standard of Pollutants for Petroleum Refining Industry' (GB 31570-2015), 'Emission Standard of Pollutants for Petroleum Chemistry Industry' (GB 31571-2015) and so on.

In addition to the national standards, various local standards have been established. For example, Beijing has established the 'Integrated Emission Standard for Air Pollutants' (DB 11/501-2007) and the 'Emission Standard of Air Pollutants for Refining and Petrochemical Industry' (DB 11/447-2007). Shanghai has newly established the 'Integrated Air Pollution Emission Standard' (DB 31/933-2015). The standards and documents established by various provinces have further strengthened the control of VOCs pollution.

Furthermore, the state has also put forward more stringent emission standards for vehicles. Since 2017, the national vehicle emission pollutants will fully implement the national V emission standards. For petrochemical and refining industries, the more stringent gasoline and diesel standards enables petrochemical enterprise to further optimize the oil processing technology and reduce the sulfur and benzene concentration in petroleum products. To a certain extent, it also can reduce the VOCs emissions from petrochemical and refining industries.

4.3 Regional Joint Prevention and Control

The complex air pollution in China is not just a local issue but a domestic regional cross-boundary problem. In order to solve this problem, regional coordination to control cross-boundary air pollutants is necessary to reduce the impact of atmospheric pollution on downwind locations, and improve the regional air quality.

Until now, nine regions, including the Beijing-Tianjin-Hebei region, the PRD region, the YRD region, central Liaoning region, Shandong Peninsula, Western Taiwan Straits Metropolitan region, Cheng-Yu Metropolitan region, Wuhan Metropolitan region, and Cheng-Zhu-Tan region, have been designated as pilot regions for the joint prevention and control of regional air pollution.

The effectiveness of regional joint prevention and control is gradually emerging. For example, on 20 August 2015, ahead of the 70th-anniversary celebrations of the end of World War II, the Beijing and surrounding areas shut down industrial facilities and reduced car emissions. This action resulted in $PM_{2.5}$ concentration lower than the national air quality standard. The restrictions resulted in an average Beijing $PM_{2.5}$ concentration of 19.5 $\mu g/m^3$, the lowest that had ever been on record in the capital, together with the Parade Blue phenomena. At the same time, various local governments (such as Beijing, Shanghai, Guangdong, etc.) are also developing different policies and plans to conduct VOCs prevention and control.

Besides, in order to improve the air quality in autumn and winter, the state has new requirements for new and existing enterprises in "2+26" cities (including Beijing, Tianjin, and 26 cities in the Beijing-Tianjin-Hebei and surrounding area). The pollutants in these enterprises will implement specific emission limits from March 1, 2018 and October 1, 2018 respectively.

4.4 VOCs Prevention Technologies

VOCs pollution prevention and control follows the principle of comprehensive prevention and control combined with source control, process control and end treatment.

Source control mainly refers to develop and use less contaminated raw materials, eliminate materials which have many types of VOCs and volatility, and use alternatives, alternative technologies. For petrochemical and refining industries, adopting advanced clean technologies, improving the conversion and utilization efficiency of crude oil are encouraged. Reducing or avoiding the emission of VOCs at the source is the primary way to control VOCs pollution. It is easy for VOCs to violate and leak during usage, transportation, and storage. Process control can be achieved by controlling leakage, collecting VOCs and so on. In addition, it also can be achieved by taking measures, such as improving the automation and sealing of equipment components, optimizing process, improving equipment, and using clean process, to reduce VOCs emissions.

End treatment is the final step to control VOCs emissions, which is the research focus and hotspots. Based on the technology principles, the end treatment techniques of VOCs can be broadly divided into two categories: decomposition and destruction technologies and recycling technologies. On the basis of investigation of VOCs pollution control technologies status, the MEE issued the 'National Advanced Pollution Prevention and Control Technology Catalogue (VOCs Prevention and Control Technology)', listing various proposed advanced VOCs end treatment technologies and its application scope (below table 5 only lists processing technologies involving the petrochemical industry). These technologies are innovative. The effects are stable and the cost is reasonable. The technologies should be encouraged for enterprises to promote applications

Table 5 Advanced VOCs end treatment technologies and its application scope

No.	Technology	Application scope
1	Activated carbon adsorption-nitrogen desorption condensation solvent recovery technology	Petrochemical and other industries
2	Oil and Gas Active Carbon Adsorption and Recovery Technology during Oil Storage and Transportation	Oil and gas recovery during product oil loading or storage
3	Oil and Gas Membrane Separation and Adsorption Recovery Technology for Oil Storage and Transportation	Petrochemical industry oil and gas recovery
4	Stationary organic waste heat storage combustion technology	Medium or high-concentration VOCs exhaust gas purification in petrochemical and other industries
5	Rotary heat storage combustion purification technology	Medium or high-concentration VOCs exhaust gas purification in petrochemical and other industries
6	Regenerative catalytic combustion (RCO) technology	Medium and high concentration VOCs exhaust gas treatment
7	Low concentration organic waste gas biological purification technology	Low concentration organic waste gas treatment
8	Advanced oxidation-biological purification coupled processing technology	Purification of low-concentration VOCs and malodorous gases from the production processes and sewage treatment plants (stations) in industries such as petroleum refining

4.5 Pollutant Permit System

China's pollutants permit system began in the 1980s, but it did not play a central role at that time. Until November 2016, the new pollutant permit system began to be implemented nationwide. It was clearly proposed that by 2020, the pollutants permit covering all stationary pollution sources should be issued, and a permit emission control system should be established, and impelling the "one-signature" management. In recent years, the state has issued 'Provisional Regulations on the Management of Emission Permits', 'Administrative Measures on Emission Permits (Trial)' and other documents to present detail provisions on the procedures for the application, review, issuance, and management of pollutant permit. At the same time, regulations on the allowable emission concentrations, permitted emissions, and environmental management for air pollutants were also provided.

For the petrochemical and refining industries, all the enterprises should be issued the pollutant permit before the end of 2018. In order to guide and regulate the application and issuance of emission permits for enterprises, the MEE issued the 'Technical Specification for Application and Issuance of Pollutant Permit in Petroleum Industry' (HJ 853-2017). At present, 34.46% of China's petrochemical and refining enterprises have completed the application of pollutant permit in the application platform. The VOC annual permit emission limit has implemented in three VOCs sources, i.e. Leakage of equipment static and dynamic sealing points, Volatilization loss of organic liquids storage and blending process and Volatilization loss of organic liquids loading and unloading process.

No pollutant emissions will be allowed without air pollution permits. One enterprise can held one permit. By implementing the pollutant permit system, it will change the phenomenon of disorderly discharge of enterprises in China, and promote the continuous improvement of environmental quality. The content of the permit license is not only to control the concentration of permitted emissions and the amount of permitted emissions, but more importantly, to control pollution prevention and control measures, finally to reduce pollutants.

Overall, from the review of the policies on the control of VOCs and/or photochemical smog, the stricter and aggressive measures help to alleviate air pollutants. Even though the atmospheric pollution is likely to be persistent because of on-going urbanization and industrialization, it is possible that the air quality can be gradually improved if appropriate measures are undertaken.

5 Problems and Suggestions

The Chinese government began to attach great importance to controlling VOCs emissions, regarding VOCs as a top priority among atmospheric pollutants for effectively improving air quality. Even though laws, regulations, standards and measures to control VOCs have been introduced in recent years, they still have problems and need to be further strengthened and standardized.

5.1 VOCs Monitoring Methods

Some standards put forward regulations and requirements to VOCs monitoring. However, the VOCs monitoring indicators and the monitoring technology methods are not clear or uniform. The characterization of VOCs is not uniform, some are based on the sum of characteristic pollutants, and some are based on NMHC. While in practice, VOCs are generally collected by solid adsorption or sampling bags, and analyzed by gas chromatography, gas chromatography/mass spectrometry, high performance liquid chromatography (HPLC) analysis, and a series of portable devices developed on this basis, such as TVA2020C, PHX-21. In addition, the pollutant permit system requires enterprises to implement self-monitoring. Now the self-monitoring of the pollutant permit has designed nearly 610,000 monitoring devices, while there are only 41077 related to the petrochemical and refining industries. Besides, the self-monitoring programs in enterprises are still based on manual monitoring, and the proportion of automatic monitoring is only 12.80%.

Therefore, it is necessary to formula and improve the VOCs monitoring standard system, establish scientific VOCs monitoring methods, and carry out research on petrochemical and industries industry monitoring method standards. In addition to monitor the conventional chimneys, leak detection standards for the devices (production devices, non-organized gas collection devices, etc.) should also be considered. Secondly, to improve monitoring capabilities, establish on-line monitoring networks for VOCs (using infrared instruments), and increase efforts to research portable monitoring devices and on-line monitoring devices. Finally, VOCs monitoring and big data platform are built to achieve standardized, precise, stable and reliable VOCs monitoring data.

5.2 VOCs Emission Inventory

The VOCs emission inventory is an indispensable parameter for the formulation of VOCs control measures. Compared with foreign countries, China's VOCs emission inventory research started late and is still on the beginning stage. There is no uniform and standardized emission source classification system, and there are few special studies on various types of emission sources. The petrochemical and refining industries have a poor level of refinement of emission source inventory, sluggish data updates, and non-comprehensive coverage. Besides, it lacks localized VOCs emission factors. The currently used emission factors are generally revised empirical emission factors that do not reflect China's actual situation. And the statistics of the activity level of the emission source sector is incomplete and lacks continuity.

Therefore, it is extremely necessary to establish an updated emission inventory that covers the spectrum of each pollution source in petrochemical and refining industries as much as possible. In addition, based on the actual situation in China, China's exemptions lists and the corresponding industry exemptions lists should be formulated. On the basis of the definition of photochemical reactivity, poisonous and harmful atmospheric pollutants shall be set up synchronously to control toxic and hazardous characteristics of VOCs. It is recommended to quote the list of HAPs formulated by the US EPA, and gradually improve the list of toxic and harmful air pollutants in China according to national conditions.

5.3 VOCs Control Technologies

The components and volumes of VOCs emitted from different processes in petrochemical and refining industries are different, and the adapted technologies are different. At present, the effect of VOCs treatment technology is different, and the non-methane hydrocarbon (NMHC) treatment efficiency is quite low. Moreover, the economic cost of some technologies is high, and many VOCs processing facilities cannot operate stably, continuously, and reliably.

Therefore, when selecting VOCs control technologies and facilities, enterprises should consider the characteristics of VOCs, as well as comprehensive consideration of various technologies' performance indicators, construction and operating costs, and implementation of emission standards. If necessary, two or more processes combination and multi-stage treatment can be used to achieve compliance with the emission standards. Special research should be conducted on VOCs in petrochemical and refining industries. Measures like continuous research and innovate treatment technologies, development relevant alternatives and improve existing equipment or processes are need to take effort to reduce VOCs emissions.

6 Conclusion

With the rapid progress of urbanization and industrialization in China, atmospheric pollution has become more and more serious. Currently, the type of atmospheric pollution is no longer a simple pollution, but a regional and complex pollution with extensive secondary pollutants. As an important precursor to cause secondary pollution, VOCs have attracted attention from countries in the world. Because of the huge VOCs emissions, wide range of sources in petrochemical and refining industries, accurately understanding the current status of VOCs emissions in the petrochemical and refining industries in China and the status of VOCs management and control are important prerequisites to improve China's ambient air quality. It is important for China's future comprehensive air pollution control decisions. Therefore, this article mainly discusses the status of China's atmospheric environment, the current status of VOCs emissions from petrochemical and refining industries, and the VOCs control measures.

China's management and control of VOCs in petrochemical and refining industries is still in its infancy. The future prevention and control of VOCs in petrochemical and refining industries remains an arduous task. In response to the current problems in the control of VOCs, some countermeasures and suggestions are proposed:

1) Formula and improve the VOCs monitoring standard system as soon as possible, and determine the characterization methods that comply with the VOCs in petrochemical and refining industries. At the same time improve the monitoring capabilities and the accuracy of online monitoring devices.

2) Accelerate the research on emission inventory and localization emission factors of VOCs in petrochemical and refining industries, establish an exemption list and a list of toxic and hazardous

pollutants, and provide technical support for the active implementation of emission inventory and source resolution of VOCs in the petrochemical industry;

3) Research and development best VOCs pollution prevention technologies for petrochemical and refining industries. Comprehensive consideration should be given when selecting VOCs treatment technologies. When necessary, a variety of technologies can be used together to achieve the pollutants meet the standards.

In the long-term perspective, the VOCs management and control model need to carry out work from the policy level, technical level, and management level. There are requirements and guidance in terms of policies, and they are technically supported and guaranteed in terms of technologies. There are breakthroughs and improvements in terms of management. Only by that, the VOCs pollution problem can effectively solved and the ambient air quality can be improved.

Acknowledgements

The work is supported by the International Cooperation Program for Innovative Talents of the China Scholarship Council (liujinwei [2016] 7988). Thanks for the funding.

References

1. Derwent R G, Jenkin M E, Saunders S M, et al. Photochemical ozone formation in north west Europe and its control[J]. Atmospheric Environment, 2003, 37(14):1983-1991.
2. Wu R, Xie S. Spatial Distribution of Ozone Formation in China Derived from Emissions of Speciated Volatile Organic Compounds[J]. Environmental Science & Technology, 2017, 51(5):2574.
3. Middleton P, Stockwell W R, Carter W P L. Aggregation and analysis of volatile organic compound emissions for regional modeling[J]. Atmospheric Environment Part A General Topics, 1990, 24(5):1107-1133.
4. Crutzen P J, Lawrence M G, Pöschl U. On the background photochemistry of tropospheric ozone[J]. Tellus Series B-chemical & Physical Meteorology, 1999, 51(1):123-146.
5. Klimont Z, Streets D G, Gupta S, et al. Anthropogenic emissions of non-methane volatile organic compounds in China[J]. Atmospheric Environment, 2002, 36(8):1309-1322.
6. Wei W, Wang S, Hao J, et al. Projection of anthropogenic volatile organic compounds (VOCs) emissions in China for the period 2010–2020[J]. Atmospheric Environment, 2011, 45(38): 6863-6871.
7. Saikawa E, Kim H, Zhong M, et al. Comparison of emissions inventories of anthropogenic air pollutants and greenhouse gases in China[J]. Atmospheric Chemistry & Physics, 2017, 17(10):1-41.
8. Huang G, Brook R, Crippa M, et al. Speciation of anthropogenic emissions of non-methane volatile organic compounds: a global gridded data set for 1970-2012[J]. Atmospheric Chemistry & Physics, 2017, 17(12):1-36.
9. Zheng J, Zheng Z, Yu Y, et al. Temporal, spatial characteristics and uncertainty of biogenic VOC emissions in the Pearl River Delta region, China[J]. Atmospheric Environment, 2010, 44(16):1960-1969.
10. Ohara T, Akimoto H, Kurokawa J, et al. An Asian emission inventory of anthropogenic emission sources for the period 1980–2020[J]. Atmospheric Chemistry & Physics, 2007, 7(16):6843-6902.
11. Xia L, Cai C, Zhu B, et al. Source apportionment of VOCs in a suburb of Nanjing, China, in autumn and winter[J]. Journal of Atmospheric Chemistry, 2014, 71(3):175-193.

12. Ying L, Min S, Lu S H, et al. Source apportionment of ambient volatile organic compounds in the Pearl River Delta, China: Part II.[J]. Atmospheric Environment, 2008, 42(25):6261-6274.
13. Rumei L I, Yuanyuan W U, Peng L, et al. Characteristics and sources apportionment of ambient volatile organic compounds(VOCs) in summer in Shuozhou[J]. Environmental Chemistry, 2017.
15. Han L, Wang X R, He M, et al. Inventory and environmental impact of VOCs emission from the typical anthropogenic sources in Sichuan province[J]. Environmental Science, 2013, 34(12):4535-4542.
16. Wallace L A, Pellizzari E D, Hartwell T D, et al. The influence of personal activities on exposure to volatile organic compounds.[J]. Environmental Research, 1989, 50(1):37.
17. Zhou J, You Y, Bai Z, et al. Health risk assessment of personal inhalation exposure to volatile organic compounds in Tianjin, China.[J]. Science of the Total Environment, 2011, 409(3):452.
18. Lin L, Jun L U, Ying-Ge M A, et al. Reviews of VOCs Control at Home and Abroad[J]. Administration & Technique of Environmental Monitoring, 2011.
19. Wang H L, Zhang G N, Nei L, et al. Study on control and management for industrial volatile organic compounds (VOCs) in China.[J]. Environmental Science, 2011, 32(12):3462.
21. Guo S, Tong L, Zhou X, et al. VOCs Emission Control in Petrochemical Industry[J]. Environmental Protection of Chemical Industry, 2014.
22. Zhang G N, Hao Z P, Jiang M, et al. Study on foreign regulations and standards of stationary sources VOCs emission control.[J]. Environmental Science, 2011, 32(12):3501.
23. Louie P K K, Ho J W K, Tsang R C W, et al. VOCs and OVOCs distribution and control policy implications in Pearl River Delta region, China[J]. Atmospheric Environment, 2013, 76(S1):125-135.
24. Zhang X, Xue Z, Sun X, et al. Review on Control of Ambient Volatile Organic Compounds (VOCs) and Suggestions in China[J]. Environmental Science & Management, 2014.
25. Khan F I, Ghoshal A K. Removal of Volatile Organic Compounds from polluted air[J]. Journal of Loss Prevention in the Process Industries, 2000, 13(6):527-545.
26. Guo H, Ling Z H, Cheng H R, et al. Tropospheric volatile organic compounds in China[J]. Science of the Total Environment, 2017, 574: 1021-1043.

513535 A Look inside a Chemical Plant Waste Water Treatment Unit

Abstract:

There are a wide variety of pieces of equipment that compose waste water treatment inside of chemical plants. In this presentation, I will discuss the process flow of the waste water treatment plant for the Syngenta St. Gabriel chemical plant to help young engineers understand the complexities of such a system and the current commercially available systems we use to treat waste water to meet government regulations. We have a pre-treatment step, carbon adsorption, and then a biological unit that help to treat the water before it can be sent out to the Mississippi River.

514113 Ethylene Risk and Technology Review: What's the Outlook?

Abstract:

Per the court-mandated timeline, the EPA is currently working through the statutorily-required risk and technology review (RTR) for the ethylene NESHAP (40 CFR, Part 63, Subpart XX and Subpart YY). Industry trade groups have been engaged in meetings with EPA as this process has moved forward. It is anticipated the proposed rule, due in late 2018, will impact current ethylene operations. It is anticipated the rule will mirror certain aspects of the refinery rule, such as the flare and PRD language and likely will address furnace stack emissions and decoke emissions (both decoke to the fire box and decoke to an atmospheric pot) and potentially waste water emissions.

This paper will discuss the possible proposed rules and the potential impact on the ethylene facilities.

AIChE PAPER NUMBER

Delivering a Successful Ethylene Section 114 Information Collection Request Test Program

David L. Elam, Jr
Project Director and Vice President
TRC

Jeff Thomason
Air Source Testing Manager
John Zink Hamworthy Combustion

Prepared for Presentation at the Spring National Meeting
Orlando, FL April 22-26, 2018

Delivering a Successful Ethylene Section 114 Information Collection Request Test Program

David L. Elam, Jr
Project Director and Vice President
TRC
Jeff Thomason
Air Source Testing Manager
John Zink Hamworthy Combustion

Abstract:

Although every emission testing project is important, some are more important than others. Routine test programs help sources demonstrate compliance with permit limits. Engineering test programs help sources make fiscally responsible decisions regarding operations or the installation of air pollution control equipment. But Information Collection Request (ICR) test programs are different – the data from these programs will be used to evaluate emission levels and may result in the establishment of emission limits or work practices that can affect the economic viability of an individual facility or an entire industry. Accordingly, ICR test programs demand the highest level of test program planning and execution to deliver defensible data that fully conforms to the requirements of the ICR.

In April 2016, EPA issued an ICR for ethylene production facilities under the statutory authority of Section 114 of the Clean Air Act (CAA). This ICR was one of the most complex that EPA has issued and presented significant technical and logistical challenges. The decoking process itself presents numerous technical challenges including elevated moisture, stack temperature variability, operational availability, and source access.

This paper and associated presentation reflect the mantra of "prepare hard, test easy." The significant technical hurdles in sampling will be addressed, as well as the logistical and personnel side of the project. When the logistical aspects of the sampling program from glassware washing procedures to the need to have 4,000 pounds of ice delivered are handled before test team deployment, the full intellectual capabilities of the test team can be focused on the range of challenges associated with field operations.

Introduction

An Information Collection Request (ICR) is a set of documents that describes reporting, recordkeeping, survey, and other information collection requirements that a federal agency imposes on the public. The Paperwork Reduction Act (PRA) requires that every federal agency must obtain approval from the Office of Management and Budget (OMB) before collecting identical or similar information from 10 or more members of the public. ICR requirements can vary in complexity, ranging from a request for existing information that is readily available to one that requires the production of substantial historical data and the generation of new data. Agencies use ICR data to comply with statutes. For example, the U.S. Environmental Protection Agency (EPA), under the statutory authority of Section 114 of the Clean Air Act uses ICRs to fulfil its requirements associated with the residual risk and technology reviews of New Source Performance Standards (NSPS) and Maximum Achievable Control Technology (MACT) standards.

A cursory search of www.regulations.gov indicates that EPA renewed or initiated close to 200 ICRs in the 12 month period beginning on July 1, 2015. Many of these were renewals, simply requesting the continued collection of data associated with previously approved ICRs; however, some ICRs imposed substantial requirements on industry. Specifically, the ICR targeting Ethylene Production Facilities issued in April 2016 certainly cost targeted facilities several hundred thousand dollars to collect and report the required information. And perhaps more important than the near-term costs associated with completing the ICR are the long-term and ongoing costs associated with regulations that will be promulgated based on ICR results.

This paper describes an approach for end-to-end management of an ICR program, focusing on specific requirements of EPA's Section 114 ICR for Ethylene Production Facilities.

Managing an ICR as a Project

Responding to an ICR is an important undertaking for an organization that should be approached using project management best practices. Although responding to an ICR can involve several hundred individual tasks and span several months, six areas require particular attention to ensure that requested information satisfies its intended purpose, is responsive, qualified, technically sound, reported accurately, and properly incorporated in rulemaking.

1. Participate in Stakeholder Meetings that Shape the ICR. EPA typically seeks involvement of the affected community when developing ICRs. Often, key trade organizations and industry groups are invited to participate in the development of the ICR. It is important to use this opportunity to understand EPA's goals in conducting the ICR, the schedule for the ICR program, and how ICR data will be used. Trade organizations and industry groups have the relevant process and operational knowledge critical to guiding the development of a practical and realistic ICR that yields the required data.

2. Assemble the ICR Response Team. An ICR response often requires efforts from environmental, operational, and information technology staff. The legal department may need to be involved to ensure protection of confidential information. Contractors and consultants may be required to support the ICR with site preparation, data collection, data analysis, reduction, and interpretation. Engage these team members early and rely on their expertise to support the ICR response through the entire process.

3. Provide Feedback on Draft ICRs. It is important to review and understand draft ICRs and comment on them in the required timeframe. Although it is easy to focus on the cost of responding to the ICR when seeking changes, the most persuasive arguments for modifying an ICR are found in the areas of safety, technical feasibility, and data defensibility. When challenging a draft ICR in any of these areas, it is important to be clear about the specific issue, the limitations of the proposed approach, and provide an alternate approach to satisfying the data need. For example, if it is unsafe to conduct a particular sampling operation, provide a defensible engineering calculation or alternative sampling approach that satisfies the data need. Similarly, if a measurement yields a detection limit that will not guide a meaningful risk assessment, explain the sampling, analytical, or process operation issues that affect the detection limit and how that detection limit will drive interpretation and use of ICR data.

4. Perform the ICR Work. Once the ICR is issued, manage the effort as a project relying on a project manager for coordinating and tracking the various tasks. Although an ICR may be issued with incremental submittal requirements, it is likely that the ICR will require simultaneous work on multiple tasks instead of independent work on sequential tasks to meet the submittal schedule. Responding to a complex ICR will require the efforts of an overall project manager and a team of task managers, who must balance ICR response with ongoing operational responsibilities.

5. Report ICR Data on Time and in the Required Format. Reported ICR data are going to be used by EPA to evaluate the effectiveness of existing regulations and determine the need for new or revised ones. It is therefore important that ICR data be collected in time to allow adequate review of the information before it is submitted. Although conformance with the ICR requirements is important, it is also important to evaluate the collected data from a broader perspective, considering data quality, completeness, and representativeness. ICRs may require electronic reporting, which can be challenging to learn. Accordingly, it is important to allow enough time for both the reporting of electronic information and the review of electronically reported information.

6. Remain Engaged in the Regulatory Process as a Stakeholder. As EPA reviews, reduces, and analyzes ICR data, it is important for affected stakeholders to understand the dataset in terms of technical accuracy and quality and remain engaged with EPA to guide the interpretation of data and the application of practical pollutant control strategies and achievable emission limits.

Overview of the Ethylene Production Facility ICR

In April 2016, EPA issued an Information Collection Request (ICR) for ethylene production facilities under the statutory authority of Section 114 of the Clean Air Act (CAA). As required by Section 112 of the statute, EPA must examine Maximum Achievable Control Technology (MACT) standards with respect to residual risk and technological advancements every eight years and promulgate revised MACT standards if the Risk and Technology Review (RTR) indicates that changes are necessary. EPA uses the authority of Section 114 of the CAA to gather information that supports the RTR evaluation process.

MACT requirements for ethylene production plants were promulgated in 2002 (40 CFR Part 63, Subpart XY; Generic Maximum Achievable Control Technology Standards and 40 CFR Part 63, Subpart XX, Ethylene Manufacturing Process Units: Heat Exchange Systems and Waste Operations). The Ethylene MACT regulates emissions of the following organic hazardous air pollutants (HAPs):

- Benzene
- 1,3-Butadiene
- Cumene
- Ethylbenzene
- Hexane
- Naphthalene
- Styrene
- Toluene
- *o*-, *m*-, and *p*-Xylene

EPA's recent Ethylene Production ICR was issued to 20 separate facilities representing a universe of 191 ethylene cracking furnaces and 37 heat exchangers. Unless a facility conducted testing that met the requirements of the ICR in the preceding five years, each facility had to select **one** ethylene cracking furnace for emission testing during cracking and decoking operations and **all** identified heat exchange systems.

The ethylene cracking furnace test programs were particularly complex, requiring multiple sophisticated sampling trains to measure the aforementioned compounds and:

- Aldehydes (Acrolein, Acetaldehyde, Formaldehyde, Propionaldehyde)
- Ammonia
- Carbon Dioxide
- Carbon Monoxide
- Ethane
- Ethylene
- Filterable Particulate Matter
- Flow Rate
- Hydrogen Cyanide
- Metal HAPs (Sb, As, Be, Cd, Cr, Co, Pb, Ni, Mn, Hg, Se, And P)
- Methane
- Oxygen
- Polycyclic Aromatic Hydrocarbons
- Propylene
- Total Hydrocarbons

The heat exchanger test programs involved collection of water samples for laboratory analysis and the use of the El Paso method to determine the following strippable compounds:

- Benzene
- 1,3-Butadiene
- Chlorine

- Chloroform
- Cumene
- Dimethylformamide
- Ethylbenzene
- Ethylene
- Hexane
- Naphthalene
- Propylene
- Styrene
- Toluene
- Total Strippable Volatile Organic Compounds (VOC)
- *o*-, *m*-, and *p*-Xylene (and total)

Although EPA identified a range of compounds that had to be measured in connection with these test programs, it did not establish detection limits for the measurements. Instead, EPA identified sample methods and run times.

EPA required the use of FTIR for several of the target pollutants. In addition, continuous emission analyzers (CEMs), two isokinetic sample trains, and Reference Method 18 sampling procedures were required to quantify the numerous target compounds. Seven (7) test runs were required for the ethylene cracking furnaces. During decoking operations, which can span several weeks due to operational schedules, EPA required 9 test runs (3 runs for each of 3 cycles). All of the test data and results had to be submitted via the Electronic Reporting Tool (ERT). Test methods and run times are summarized in Tables 1 and 2.

Table 1: Furnace Stack - Test Matrix				
Parameter	**Test Method**	**# of Test Runs**	**Run Length**	**Units of Measure**
Aldehydes[1]	ASTM D6348-03	7	60 minutes	lb/hr and ppmvd @ 3% O_2
Ammonia	ASTM D6348-03	7	60 minutes	lb/hr and ppmvd @ 3% O_2
Organic HAP[2] (1,3-Butadiene[3])	EPA Method 18 (ASTM D-6348-03)	7	60 minutes	lb/hr and ppmvd @ 3% O_2
Polycyclic Aromatic Hydrocarbons (PAH)[4]	CARB Method 429	7	Minimum of 2 dry cubic meters (~2 hours)	lb/hr and µg/dscm @ 3% O_2
Total Hydrocarbons (THC)	EPA Method 25A	7	60 minutes	lb/hr and ppmvd @ 3% O_2
Methane Ethane Ethylene Propylene	ASTM D6348-03	7	60 minutes	lb/hr and ppmvd @ 3% O_2
Nitrogen Oxides (NOx)	EPA Method 7E	7	120 minutes, minimum	lb/hr and ppmvd @ 3% O_2
Filterable PM and Metal HAP[5]	EPA Method 5 EPA Method 29	7	Minimum of 4 dry cubic meters (~4 hours)	lb/hr and µg/dscm @ 3% O_2
Hydrogen Cyanide	ASTM D6348	7	60 minutes	lb/hr and µg/dscm @ 3% O_2
Carbon Monoxide	EPA Method 10	Concurrent with all test runs	4 hours	lb/hr and ppmvd @ 3% O_2
Flow Rate[6]	EPA Method 2 O_2-EPA Method 3A EPA Method 4 (CO_2-ASTM 6348-03)	Concurrent with all test runs (7)	4 hours (60 minutes)	

[1]Acrolein, Acetaldehyde, Formaldehyde, Propionaldehyde
[2]Benzene, Cumene, Ethylbenzene, Hexane, Styrene, Toluene, o-Xylene, m-Xylene, p-Xylene
[3]1,3-Butadiene measured by ASTM D6348-03
[4]The target analyte list is included as Table 1 of CARB Method 429
[5]Metal HAP includes compounds of Antimony, Arsenic, Beryllium, Cadmium, Chromium, Cobalt, Lead, Nickel, Manganese, Mercury, and Selenium and emissions of Phosphorus. Report concentrations and mass emissions rates of each pollutant separately.
[6]Measured in conjunction with M5/29

Table 2: Decoking Pot Stack - Test Matrix				
Parameter	**Test Method**	**# of Test Runs**	**Run Length**	**Units of Measure**
Aldehydes[1]	ASTM D6348-03	3 runs per cycle	60 minutes	lb/hr and ppmvd @ 3% O_2
Ammonia	ASTM D6348-03	7	60 minutes	lb/hr and ppmvd @ 3% O_2
Organic HAP[2] (1,3-Butadiene[3])	EPA Method 18 (ASTM D6348-03)	3 runs per cycle	60 minutes	lb/hr and ppmvd @ 3% O_2
Polycyclic Aromatic Hydrocarbons (PAH)[4]	CARB Method 429	3 runs per cycle	Minimum of 2 nominal actual cubic meters (~2 hours)	lb/hr and µg/dscm @ 3% O_2
Total Hydrocarbons (THC)	EPA Method 25A	3 runs per cycle	60 minutes	lb/hr and ppmvd @ 3% O_2
Methane Ethane Ethylene Propylene	ASTM D6348-03	3 runs per cycle	60 minutes	lb/hr and ppmvd @ 3% O_2
Nitrogen Oxides (NOx)	EPA Method 7E	3 runs per cycle	120 Minutes, Minimum	lb/hr and ppmvd @ 3% O_2
Filterable PM and Metal HAP[5]	EPA Method 5 EPA Method 29	3 runs per cycle	Minimum of 4 nominal actual cubic meters (~4 hours)	lb/hr and µg/dscm @ 3% O_2
Hydrogen Cyanide	ASTM D6348	3 runs per cycle	60 minutes	lb/hr and µg/dscm @ 3% O_2
Carbon Monoxide	EPA Method 10	Concurrent with all test runs	4 hours	lb/hr and ppmvd @ 3% O_2
Flow Rate[6]	EPA Method 2 O2-EPA Method 3A EPA Method 4 (CO2-ASTM 6348-03)	Concurrent with all test runs (7)	4 hours (60 minutes)	

[1]Acrolein, Acetaldehyde, Formaldehyde, Propionaldehyde
[2]Benzene, Cumene, Ethylbenzene, Hexane, Styrene, Toluene, o-Xylene, m-Xylene, p-Xylene
[3]1,3-Butadiene measured by ASTM D6348-03
[4]The target analyte list is included as Table 1 of CARB Method 429
[5]Metal HAP includes compounds of Antimony, Arsenic, Beryllium, Cadmium, Chromium, Cobalt, Lead, Nickel, Manganese, Mercury, and Selenium and emissions of Phosphorus. Report concentrations and mass emissions rates of each pollutant separately.
[6]Measured in conjunction with M5/29

EPA established an aggressive schedule for ICR completion as indicated in the following table:

Table 3: ICR Schedule	
ICR Element	**Due Date**
Submit Historical Stack Testing Results for Ethylene Cracking Furnaces, Decoking Operations, and Heat Exchanger Systems	July 29, 2016
Submit Heat Exchanger System Sampling and Analysis Data	September 30, 2016
Submit Ethylene Cracking Furnace Stack Testing Emissions Data	November 30, 2016
Submit Decoking Operations Stack Testing Emissions Data	November 30, 2016

EPA did not require affected facilities to submit test plans in connection with these test programs. Although test plans were not required, TRC developed test plans for ethylene production ICR test programs that conformed to the requirements of ASTM D7036 (Standard Practice of Competence for Air Emission Testing Bodies). Test plans are critically important for complex test programs that require large test teams, multiple and simultaneous test methods, specific unit operating conditions, the collection of defined operational data, and specific reporting requirements. Documenting test program requirements in advance of the test program allows the facility and test team to anticipate problems, develop contingency plans, and ensures that all ICR program requirements are satisfied.

Execution of the Ethylene Production Facility ICR

The ethylene production facility ICR test program presented safety, scheduling, staffing, method application challenges, logistical support, and reporting. This section highlights those challenges and TRC's approach to managing them.

- **Safety:** All emission testing work within a petrochemical facility requires heightened attention to safety. Safety management for the ethylene production facility ICR test program was especially challenging due to the heat at the sample location (120°F), the size of the test team (8-9 people), the crowding associated with multiple sample trains, and the 14-16-hour work days (in some cases, round-the-clock testing in two shifts). TRC assigned a dedicated to safety officer to the program, selected a test team with previous petrochemical experience, mandated hourly rest and hydration breaks in an air-conditioned area, and scheduled the test program over multiple mobilizations to satisfy both test program requirements and to provide a break for team members. In addition to the onsite safety issues, managing the front end safety processes associated with Transportation Worker Identification Credential (TWIC) requirements, background screening, employee drug testing, and site-specific requirements was challenging.
- **Scheduling and Staffing:** Each facility had to select one ethylene cracking furnace for emission testing during cracking and decoking operations and all identified heat exchange

systems. As directed in the ICR, three (3) runs had to be completed on the furnace stack and decoking pot stack simultaneously during three separate decoking cycles, for a total of eighteen (18) runs (9 runs per source) for all parameters listed in Table 1 (furnace stack) and Table 2 (decoke pot stack). To meet operational and personnel scheduling requirements, TRC completed the test programs in four separate mobilizations. The first mobilization lasted 2 days and required 3 people to complete testing of the heat exchange system. The second mobilization lasted 10 days and required 8 people to complete 7 furnace runs and 3 decoking runs. The third and fourth mobilizations each lasted 4 days and required 9 people to complete 3 decoking runs. Careful scheduling was required to coordinate operational requirements, assign testing team staff with petrochemical experience using complex and sophisticated test methods, and to ensure continuity of the test team for the four separate mobilizations.

- **Test Method Application:** The ICR identified 13 required test methods and provided alternates for 5 of the required test methods. Test method selection had to simultaneously satisfy source access, simultaneous testing requirements, and detection limits. After reviewing the list of compounds for which Method 320 was required, TRC opted for FTIR analysis by alternate ASTM Method D-6348. When Method 18 analysis was required and Method 320 and ASTM D-6348 were allowed alternates, TRC opted for the ASTM Method D6348 for only 1,3-butadiene. This approach allowed us to apply the robust quality assurance procedures in ASTM D-6348 where FTIR is the analytical method of choice and to apply conventional gas chromatography with flame ionization for hydrocarbon HAPs.
- **Logistical Support:** Each test program mobilization required considerable logistical support. Travel arrangements were required for test personnel coming from different TRC offices. Expendable supplies, particularly ice for the sampling trains and liquid nitrogen for the FTIR, had to be available onsite in sufficient quantities to support the test program. Additionally, couriers and offsite laboratory support were required to analytical requirements for filterable particulate matter, polyaromatic hydrocarbons, and metals.
- **Data Reduction and Reporting:** The ICR test program generated sampling and analytical data streams from a number for sources including handwritten physical measurements, continuous emission monitoring data, analytical data from the onsite FTIR and gas chromatograph, and analytical data from subcontract laboratories. Data reduction was labor-intensive requiring multiple quality control reviews to ensure accuracy of transcribed and merged electronic data. The final report for one test program spanned 3,254 pages. Importantly, the data had to be reported using EPA's Electronic Reporting Tool, a time-consuming process that is most efficiently performed when all data are available.

Although the ICR did not require a formal test plan, TRC believes that the successful completion of these test programs was accomplished by attentive test program planning. Detailed test planning, performed by an experienced project manager, ensured that the ICR test programs were completed safely by experienced personnel, without delays due to logistical support.

AIChE Paper Number: 165c

Predicting Long Term Evaporative Cooling Tower Water Usage: The Energy-Water Balance

William Celenza

Technical Professional Leader – Process

KBR, Inc.

Sarah Shank

Sr. Technical Professional – Process

KBR, Inc.

Prepared for Presentation at the 2018 Spring National Meeting
Orlando, Florida, April 25, 2018

Predicting Long Term Evaporative Cooling Tower Water Usage: The Energy-Water Balance

William Celenza
Technical Professional Leader – Process
KBR, Inc.

Sarah Shank
Sr. Technical Professional – Process
KBR, Inc.

Abstract:

Evaporative Cooling Towers are commonly used to dissipate heat from industrial process operations by cooling water through a combination of heat and mass transfer. An evaporative cooling tower can operate over a wide range of recirculating water rates, air velocity rates, and heat loads with variation in the approach of cold water to the ambient wet-bulb air temperature. Designed to manage summer worst case ambient conditions, the actual long term water and/or energy usage is overestimated, increasingly so for locations which experience extremes in seasonal weather. For instance, some evaporative cooling systems managing a constant industrial heat load can cool over 50% by non-evaporative air-cooled heat transfer; during reduced ambient air temperatures.

This paper presents a procedure for predicting long term water usage of a cooling system based on regional climactic data. The analysis focuses on industrial cooling systems consisting of wet towers that supply a common set of process loop heat exchangers, typically arranged in parallel; and will provide analysis for water or energy minimization. The estimated water usage on a typical monthly, seasonal or annual basis is presented to demonstrate the energy –water tradeoff for sites across the continental United States. The results estimate how much operational flexibility is available for a cooling system and offers an understanding of the interaction between water and energy requirements under varying climate conditions.

Introduction

Cooling towers are used for heat exchange, and when used to extract heat from operations in the oil refining and chemical process industries are often referred to as industrial process cooling towers. The design of cooling towers is based on a specified heat load, which in the ethylene industry is based on production and thus relatively constant throughout the year.

Water is the preferred heat-transfer medium because of availability, high heat capacity, and relatively low costs. Wet cooling towers dissipate the heat by recirculating water in direct contact with atmospheric air through convection and evaporation of water; measured as sensible and latent heat transfer, respectively. Both of these mechanisms operate at the air-water boundary. The driving force for heat and mass transfer is taken as difference in enthalpy between the water and air; although theoretically speaking the difference in water vapor pressure between the air and water phases is the actual driving force (1).

Figure 1 Typical Wet Cooling Tower System Arrangement

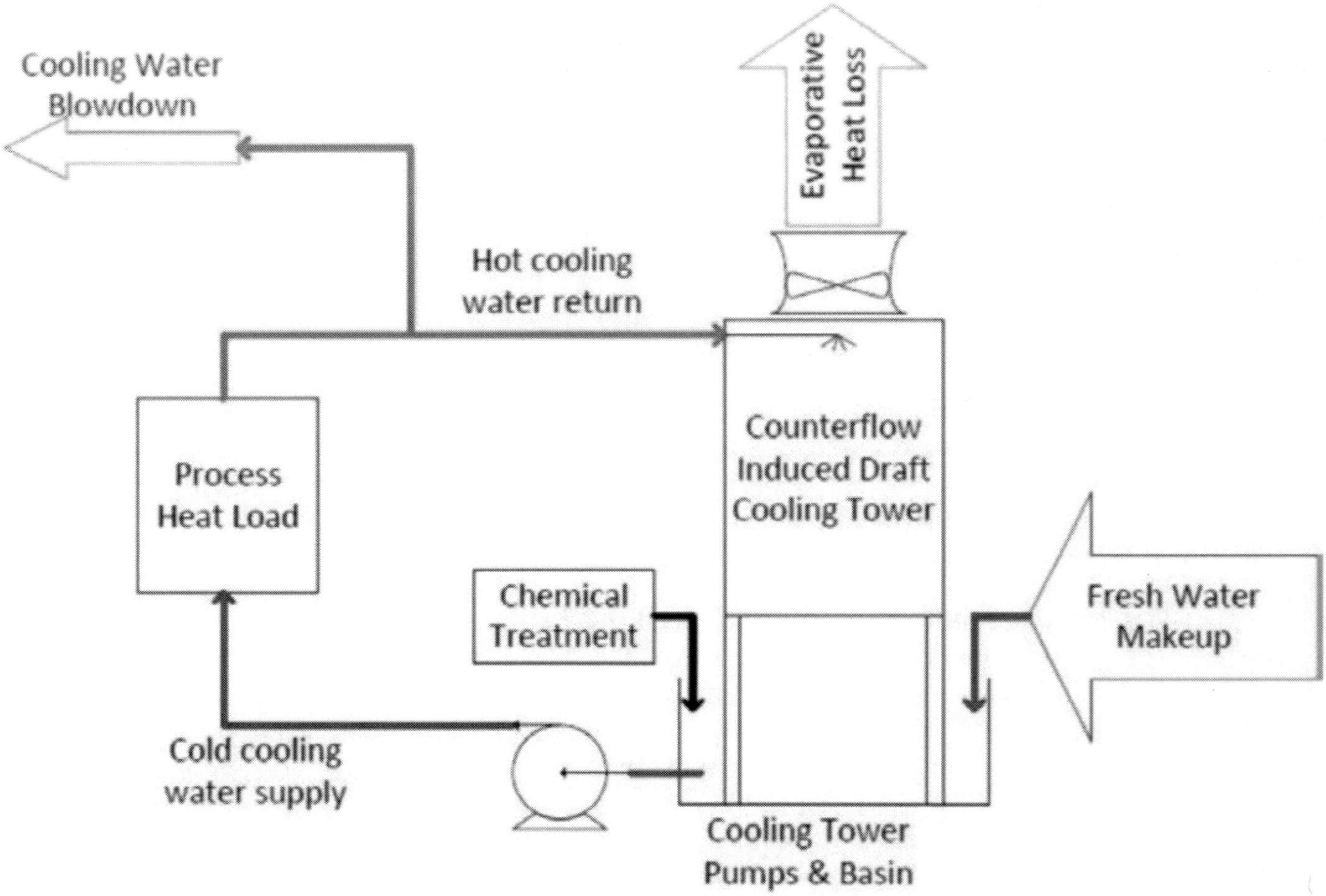

For maintenance reasons, heat transfer surfaces, such as heat exchanger tube walls, require a restriction be placed on cooling water velocity to prevent corrosion and fouling. Thus the combination of flow velocity and process heat load dictates the range, or fixed temperature reduction, the cooling tower's recirculating water must accommodate. A typical Wet Cooling Tower System Arrangement is depicted in Figure 1.

Open circuit counterflow-induced draft cooling towers, as shown in Figure 2, are predominant in the process industries; and only this type is considered further to simplify this analysis.

Figure 2 Typical Counterflow Induced Draft Cooling Tower Arrangement

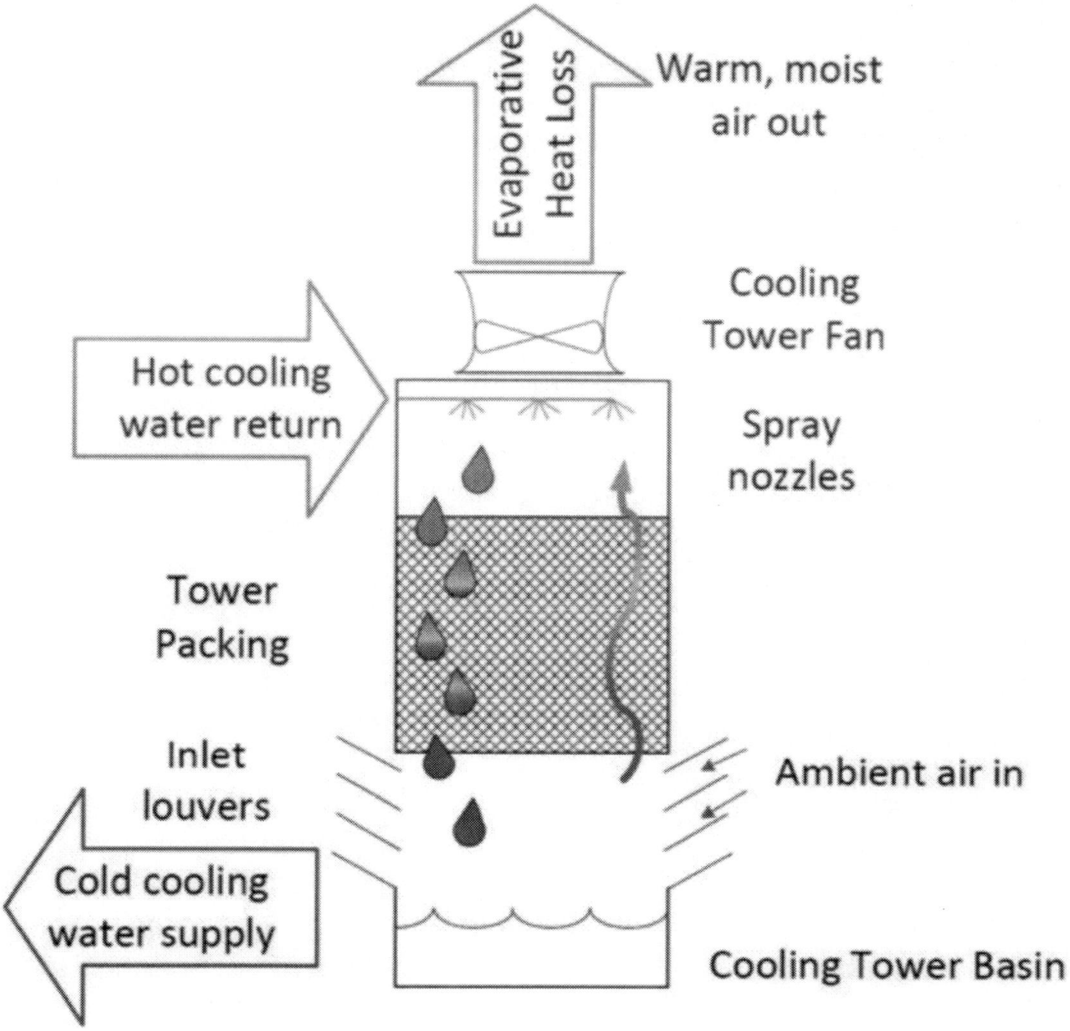

Cooling tower heat and mass transfer is based on evaporative cooling to the atmosphere. The maximum efficiency is limited by the local air temperature and ambient humidity, collectively measured as the wet-bulb temperature. The meteorological quantity called the wet-bulb temperature is measured with a normal thermometer that has a damp cloth covering the bulb which is fully ventilated. The wet-bulb temperature is always lower than the usually reported "dry-bulb" temperature; with how much lower dependent on the humidity. At 100% humidity the dry-bulb and wet-bulb temperatures are the same. Even though the amount of heat that needs to be extracted from the process is independent of the weather; the design and operation of wet cooling towers is dependent on ambient weather conditions. The design of wet cooling towers needs to ensure adequate cooling for continuous production under expected atmospheric conditions. However 99% of the time, cooling towers are not called upon to evaporate water at the design rate; creating an opportunity to conserve energy or water resources much of the year.

Predicting water usage in cooling systems can serve as a tool to provide for more efficient water resource management. Lakovic (2) reports that over 60% of the world's reliance on water is due to cooling purposes; and that by applying a re-circulation cooling system, such as a wet cooling tower, up to 98% savings in water consumption can be realized over once-through cooling (2,3,4). Though the design of wet cooling towers is a competitive field, where models and constants can be proprietary an approximate design is presented here using rules-of-thumb to provide for water evaporation rates and air blower power requirements. For this paper, the ethylene production facility wet cooling tower design outlined in Table 1 was used as a base case for a simplified model to predict water usage in wet cooling towers.

Table 1 Base case ethylene production facility wet cooling tower design

Parameter	Value	Comment
Ethylene Production Rate	1,000 MM lb/yr	Instantaneous
	890 MM lb/yr	Nameplate
Cooling Tower Water Recirculation Rate	105,000 gpm	Considered to be constant
Cooling Tower Design Temperature Range	15.8 F degrees	Design Hot Water Temp = 100 F, Design Cold Water Temp = 84.2
Design Outlet Air Temperature	92.1 F	Outlet Air temp = (Hot Water Temp + Cold Water Temp)/ 2 For more information see reference (5).
Design Liquid / Gas (Water to Air) ratio	1.1	L/G ratio = (Enthalpy Outlet Air - Enthalpy Inlet Air) / (Specific Heat * (Hot Water Temp – Cold Water Temp)). For more information see reference (5).
Ambient wet-bulb temperature increase to account for drift recirculation	1 F degree	Used as the cooling tower design wet-bulb temperature.

As tracking water consumption is just one piece of a water management program, this paper also presents how an energy management strategy can be applied to water conservation. The discussion first presents a water use analysis, then identifies key components of an energy strategy that can be applied to wet cooling tower water management.

Predicting Water Usage

Water is used in two ways in the ethylene production process: indirectly for heat transfer as cooling water or steam, and directly in the process. Most of the water is used for steam production and cooling towers, with up to 75% routinely used to replace water lost to evaporation in cooling towers (3). The remaining 25% is used to supply both process and other utilities, such as replacing cooling tower blowdown. Less than 10% of the supplied water sees a direct process use, and the supply water used as cooling tower makeup which is the sum of the evaporation, blowdown, and drift rates can account for 90% of water use in ethylene production. Note also that cooling tower blowdown can contribute over 75% of the total process wastewater, or dry weather flow.

Table 2 Summer and winter weather design conditions for evaporative cooling towers at nine US locations (6b, 6c)

	Ambient Air Temperature (F)				Cooling Water Cold Temperature			
Location	**Summer Wet Bulb, (a)**	**Summer Dry Bulb, (b)**	**Winter Wet Bulb, (a)**	**Winter Dry Bulb, (b)**	**Summer Supply F**	**Average Monthly Supply F**	**Winter Supply F**	**Max Season Variation %**
New Orleans, LA	81	89	31	34	84	74	55	35
Brownsville, TX	80	88	38	40	84	77	60	29
Houston, TX	79	88	29	31	84	75	55	35
Belleville, IL	78	88	8	9	84	68	43	49
Des Moines, IA	77	88	-4	-3	84	65	37	57
Phoenix, AZ	75	97	35	40	84	74	61	28
Pittsburgh, PA	73	82	6	7	84	69	45	47
Minot, ND	70	85	-18	-18	84	64	33	61
San Francisco, CA	63	74	38	40	84	78	70	17
Average	**75**	**87**	**18**	**20**	**84**	**71**	**51**	**40**

Notes:

a) For Summer Design Condition, the Wet Bulb Temp is 1.0% Occurrence Value. For Winter Design Conditions the Wet Bulb Temp is the Mean Coincident (Average) Value at the 99.0% Dry Bulb Temp Occurrence (6c).
b) For Summer Design Condition, the Dry Bulb Temp is the Mean Coincident (Average) Value at the 1.0% Wet Bulb Temp. For the Winter Design Condition, the Dry Bulb Temp is the 99.0% Occurrence (6c).

Summer and winter weather design conditions for evaporative cooling towers at nine US sites that are near to ethylene production facilities are included in Table 2. Design wet-bulb temperatures usually vary from 60 to 85 F in the US depending on the geographic area, and are usually equaled or exceeded only 2.5 percent to 5 percent of the total summer operating time (7). In the United States summer was standardized as June through September. However, the design values listed in Table 2 are annual values not seasonal values; with the 1% annual value usually about a degree cooler than the 2.5% seasonal value. The 99% winter design annual values are generally cooler than the 99% seasonal values that are based on the three coldest months of the year: December through February (6a). Recognizing that engineered solutions are seldom designed for the worst possible natural event; for petrochemical applications, a stringent 1% wet bulb exceedance level has been considered appropriate (8). Although not covered in this analysis, cooling tower performance during exceedance times is a concern for uninterrupted industry operation.

For development of seasonal or annual water usage patterns, mean monthly average wet bulb temperatures were collected for each site and used to model cooling tower operations. The mean monthly average wet-bulb temperature was derived from the mean monthly average dry-bulb temperature and the mean monthly average humidity ratio (6b). The cooling tower industry typically quotes the cooling in a cooling tower is on average 75% through an exchange of the latent heat of evaporation of water (5), with the balance through the sensible heat loss from the water. As further stated in reference 17, the percentage of heat removed by evaporative cooling in wet cooling towers varies between 60% and 90% during typical operating and ambient conditions. Analysis of the data in Figure 3 also indicates that on a yearly average basis across all the sites studied, 75% of the heat transfer is due to evaporative cooling.

Both forms of transferred heat together raise the dry-bulb and wet-bulb temperature of the air passed through the tower, which is then exhausted to the atmosphere (9). However, for the same heat load the ratio of latent to sensible heat transfer will vary over changing ambient air conditions. Figure 3 shows how the percentage of total heat transfer in a cooling tower due to water evaporation or latent heat can change during the months of the year. For most locations in Figure 3, the percentage varies from 60% to 90% over the course of the year; however, the total variation is from 40% to 105%. Note that under certain conditions, such as summer in Phoenix, both the water and air are cooled by evaporation so the evaporative cooling exceeds 100% of the cooling water effect measured only on the water.

The wet cooling tower size is a function of the water recirculation flow rate, the temperature change of the recirculating water, the wet bulb temperature, the approach to the wet bulb temperature of the returned cooled water, the air velocity through the tower, and heat transfer area provided by the tower dimensions with fill pack, if any. The hot water temperature minus cold water temperature is termed the cooling range and the difference between cold water and wet-bulb temperature is called approach. For a given cooling tower design, the quantity of water to be cooled and its inlet temperature is usually known. The outlet water temperature is specified, so the design range is set. With a design ambient wet-bulb temperature dictated by site location, only an approach would need to be chosen. In design practice the actual temperature of the cooled water is higher than the wet bulb temperature of the entering air usually by 10 to 15 degrees Fahrenheit, but seldom less than five degrees Fahrenheit.

Figure 3 Seasonal Variation of Percentage of Total Cooling Tower Heat Transfer Due to Latent Heat (Evaporative Cooling)

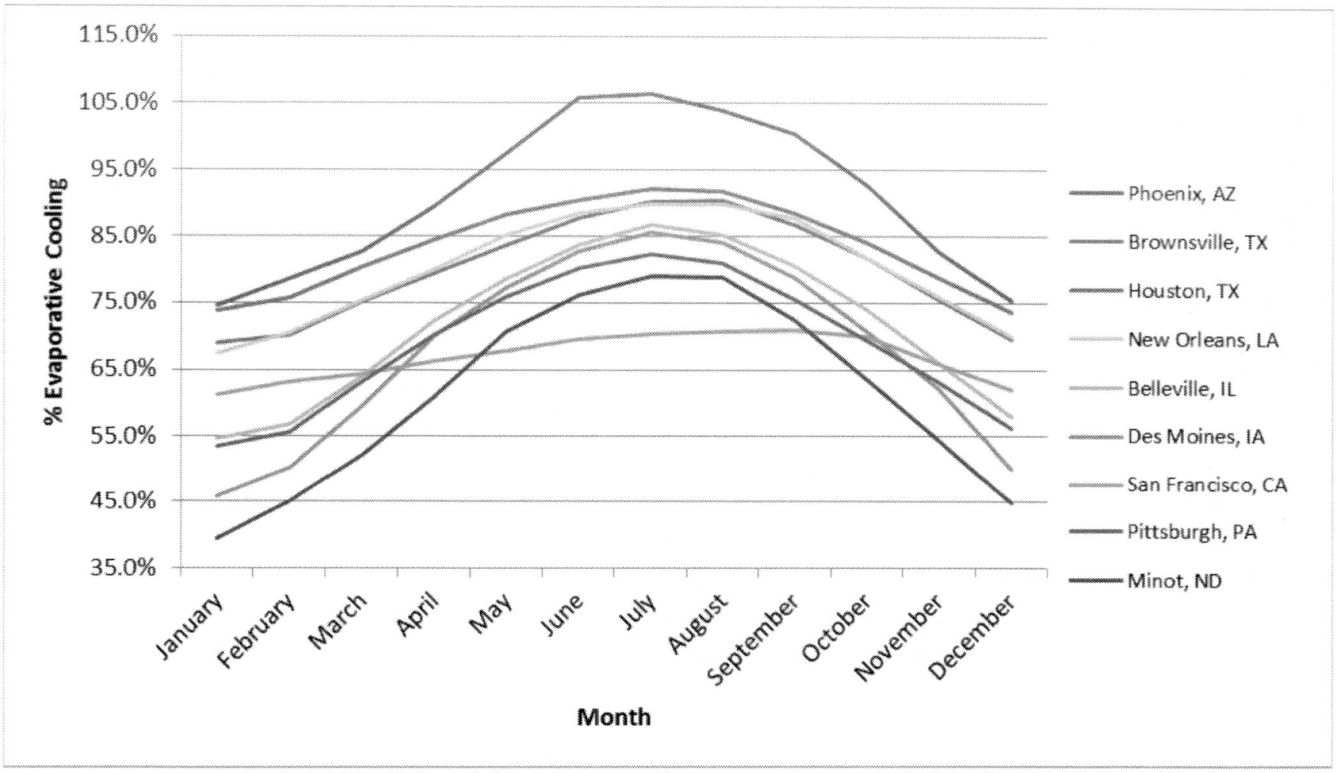

In turn for non-design ambient conditions, the size of the cooling tower, the recirculating water flow rate and the actual ambient wet-bulb temperature determine the inlet and outlet water temperatures, but not the difference between them. The amount of heat transferred to the atmosphere by the cooling tower is always equal to the heat load imposed on the tower, while the temperature at which the heat is transferred is determined by the thermal capability of the cooling tower and the entering wet-bulb air temperature. Increased tower performance can only be achieved by adding surface area or by boosting the air flow (10).

As stated earlier, cooling tower performance is related to ambient wet-bulb temperature. Wet bulb temperature is a site condition. At a constant ambient temperature, the lower the wet bulb, the drier the air, the more moisture the air can absorb and the more latent heat a given cooling tower can transfer. The highest wet-bulb temperatures occur during summer months, when air temperatures and humidity are the highest. During all other times, the cooling tower is likely oversized for the actual required heat transfer; and the cold water temperature will simply be less than the design supply temperature. Thus the cooling water temperature will vary by the season of the year, with all nine locations experiencing an average 40% difference between summer and winter design temperatures as shown in Table 2; even with modulating air flow to at least partially compensate for lower ambient temperatures. For this analysis the design cooling water supply temperature has been taken as 84 F as listed in Table 2, with the resulting monthly average supply temperature across all the sites being 71 F. The monthly average temperatures themselves can vary over 30 degrees F during the year, with an average site monthly variation of 20 degrees F during the year.

Note that in operation of cooling tower systems during subfreezing weather, a by-pass circulation method may be used to protect the tower from icing. For instance if the cooling water supply temperature should fall below 40F, a partial or full by-pass of the hot process cooling water may be implemented. In addition, induced draft fans may be reversed periodically in order to de-ice the air intakes. Thus the theoretical benefit of the ambient conditions towards process cooling may not be realized in practice.

Appreciating how seasonal design conditions influence the cooling tower operation is important to understand. Hot and cold water temperatures fall when the ambient wet-bulb temperature falls at constant heat load, range, and water flow rate. The cold water temperature from the cooling tower decreases in a linear relationship, reported as between 1.5 to nearly 2 degree wet-bulb temperature change resulting in a 1 degree cold water temperature change (5, 9, 11). This relationship is also a weaker function of the recirculating water to air ratio. For this analysis, a 1.7 ratio of the change in the wet-bulb temperature to the cold cooling water temperature is used and assumed constant over varying water to air ratios of 1 to 1.25. Thus when the wet-bulb temperature changes from 81F to 31F, or 50 degrees, from summer to winter for New Orleans as listed in Table 2 the predicted approach, the difference between the wet-bulb and cold water temperatures, would have changed only 50/1.7 or 29 degrees. The resulting cold water temperature would then be 84 – 29 = 55F at the 31F wet-bulb temperature.

However, the air's overall ability to absorb moisture reduces with temperature; the maximum absolute humidity level is halved for every 20 degree Fahrenheit temperature drop. Air at a higher wet-bulb temperature is capable of absorbing more heat according to the psychrometric properties of air. Therefore, at the lower wet-bulb temperatures a greater proportion of heat rejection is sensible; so the amount of water evaporated is reduced compared to the summer design condition (11). Figure 4 shows the predicted change in water lost to evaporation over the course of a year at the different locations. According to the ASHRAE (12) rule of thumb, the evaporation rate at typical summer design conditions is approximately 1% of the water flow rate for each 12.6F of water temperature range. Per the ASHRAE rule of thumb, the 15.8 F cooling water range listed in Table 1 for this analysis would require a 1.25% recirculating water evaporation rate. Although most sites in Figure 4 are within a 1.2 to 1.4% band during the summer, all are below 1.2% in the winter months and half are below 1% during winter. By analysis of Figure 4, the average monthly predicted evaporation rate for all sites over the year is 1.1%. This predicted rate is less than the ASHRAE predicted evaporation rate because the sensible component of the total heat transfer increases as the entering air bulb temperature decreases (4). The ratio of latent heat to sensible heat is important in analyzing water usage of a cooling tower. The ASHRAE rule of thumb appears to be more accurate for summer conditions, although even the Summer Design Conditions listed in Table 2 are predicted to average a 1.4% circulating water evaporation rate. For comparison, the Winter Design Conditions listed in Table 2 are predicted to average a 0.7% circulating water evaporation rate.

Figure 4 Seasonal Variation in Percentage of Recirculating Water Lost to Evaporation for Constant Cooling Load

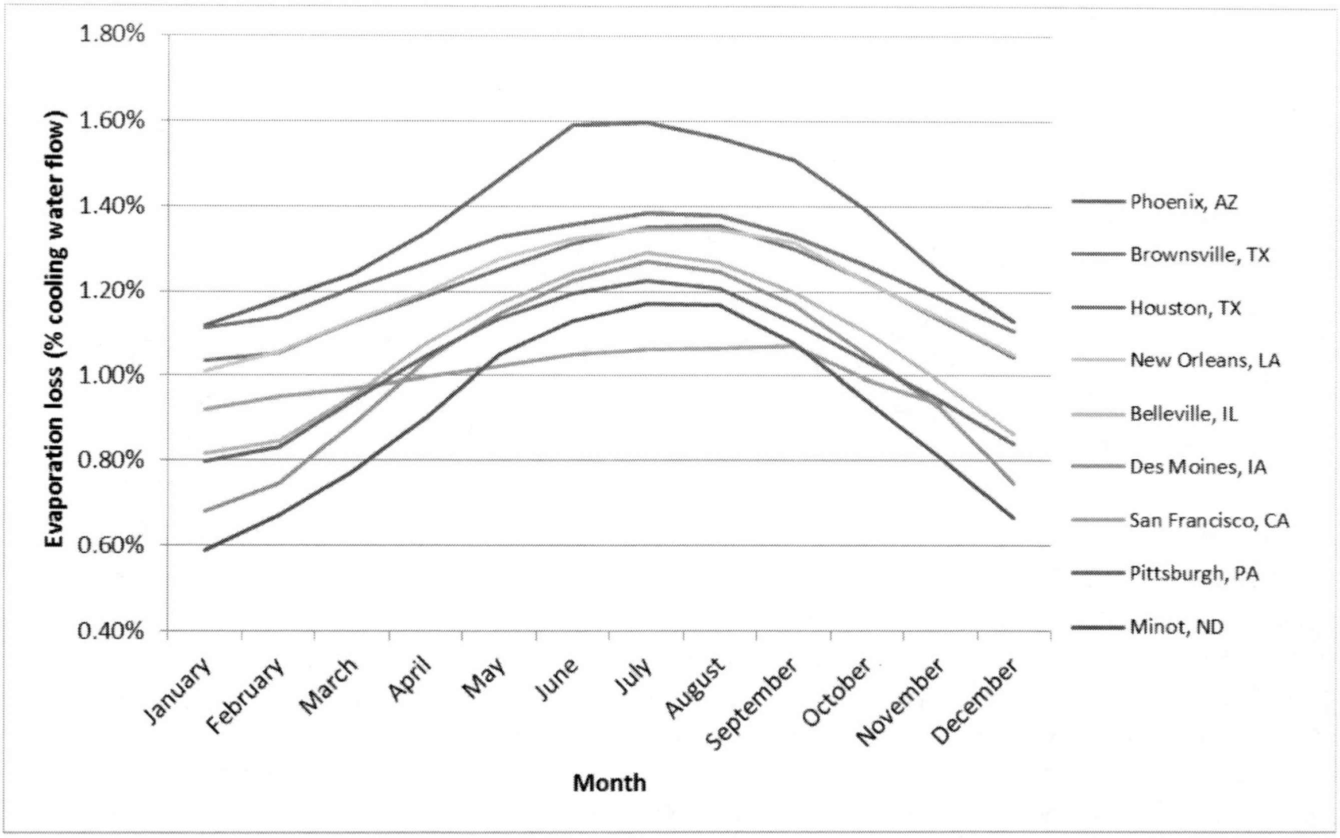

Based on the data presented in Table 3, average July summer conditions for a cooling tower in Pittsburgh, PA would be predicted to require 6% less water usage than at the Summer Design Conditions. For average January winter conditions, a cooling tower in New Orleans, LA would be predicted to require 30% less water usage than at the Summer Design Conditions. On an annual basis, across all these locations a cooling tower would be predicted to use on average 20% less water than under the Summer Design Conditions.

Table 3 Relative predicted evaporative cooling water loss vs. Summer Design Condition at nine US locations for the same cooling tower design

Location	Summer Design (% Evaporative Water Loss)	Relative Predicted Evaporative Water Loss (% difference from Summer Design)			
		Average July	Average January	Winter Design	Annual
New Orleans, LA	1.43	-6	-30	-40	-16
Brownsville, TX	1.42	-2	-22	-36	-11
Houston, TX	1.41	-4	-27	-41	-15
Belleville, IL	1.4	-8	-42	-57	-24
Des Moines, IA	1.4	-9	-51	-63	-28
Phoenix, AZ	1.6	0	-30	-41	-15
Pittsburgh, PA	1.31	-6	-39	-52	-21
Minot, ND	1.36	-14	-57	-73	-33
San Francisco, CA	1.13	-11	-23	-30	-16
Average	**1.4**	**-7**	**-35**	**-48**	**-20**

In addition, as the evaporation rate is lowered by site ambient weather conditions; related site water usage would also be lowered. For instance, the blowdown flowrate is a function of the cycles of concentration in the cooling tower; which is dependent on the quality of the cooling tower makeup water supply. The blowdown is related to the evaporation rate through the cycles of concentration. At cycles of 4 to 5, the incoming makeup water quality would be concentrated at 4 to 5 times supply concentration in the recirculating cooling water by control of the blowdown rate. When the evaporation rate reduces 10%, the blowdown rate would reduce 3% for a cycle of 4, and 2.5% for a cycle of 5. The utility water use and wastewater generation flows then would reduce in direct proportion that the blowdown represents these flows. The blowdown is usually less than 15% of total utility use and up to 75% of process or dry weather wastewater flows at an ethylene production facility. So in comparison, a 10% reduction in the evaporation rate could reduce the water supply demand up to 7.5% in addition to 0.3% from blowdown; however, the evaporation impact on the wastewater flows would be limited to only the reduction in the required blowdown rate, which could translate into up to 2% less generated wastewater.

Note that the values presented in Figure 4 and Table 3 assume that a cooling tower can effectively modulate the air blower flow rate to adjust cooling tower capacity to accommodate ambient conditions. The correlation of energy usage and water usage is discussed in the next section.

Energy Management

Since cooling is primarily by transfer of latent heat, cooling tower size is based on the total heat content or enthalpy of the entering air. At any one enthalpy condition, the wet-bulb temperature is constant. Thus cooling towers are selected and guaranteed to cool a specific volume of water from a hot water temperature to a cold water temperature while operating at a design wet-bulb temperature, or constant enthalpy. However, most cooling towers encounter substantial changes in ambient wet-bulb temperature during the year. A cooling tower design should provide for energy efficient operation during the majority of the year when atmospheric conditions do not approach design cases and the tower has excess capacity.

The major operating expense of a cooling tower is the fan power consumption (5), which can be balanced against water usage partly by design choices in the tower construction. A low air rate requires a large tower, while a high rate in a smaller tower requires greater fan power. Typical limits in air velocity are 300 to 700 fpm in counterflow towers (4). The Liquid/Gas (L/G) ratio of a cooling tower is the ratio between the water and the air mass flow rates. A selected L/G ratio for tower operation depends on the type of fill media, but ranges from 1 to 2.

Cooling rates of mechanical draft towers depend upon their fan diameter and speed of operation. To maintain energy efficient operation at lower ambient temperatures, reducing airflow would be an option to reduce capacity; with fan speed reduction usually desirable over closing dampers on air intakes which just reduce fan housing efficiency. The capacity reduction can be achieved with fan cycling, multi-speed motors, extra motors, or variable speed or frequency drives. Fan cycling, or turning off the motor of one or more of a series of fans, is an easy method but does not work well when close temperature control is required and can result in frequent motor starts (10).

Table 4 Cooling tower capacity, fan speed and energy (13)

Fan Speed	Tower Capacity	Energy Use
100%	100%	$1^3 = 1 \times 1 \times 1 = 100\%$ hp
90%	90%	$0.9^3 = 0.9 \times 0.9 \times 0.9 = 72.9\%$ hp
80%	80%	$0.8^3 = 0.8 \times 0.8 \times 0.8 = 51.2\%$ hp

As shown in Table 4, Cooling Tower capacity varies directly with the fan speed, but energy consumption varies with the cube of the speed ratio. Thus reducing the fan speed to 80% would provide for a nearly 50% reduction in horsepower demand. For instance, variable frequency drives coupled with a standard fixed-pitch fan would provide the ability to modulate fan motor speed to provide capacity control along with energy savings (4). At a minimum, two-speed fans are typically recommended since the high and low speeds also allows for flexibility in controlling the return water temperature. For two-speed fans with single winding motors, the low speed is always half of full speed (10). However, for industrial facilities with constant heat loads, the change in ambient conditions usually does not require a compensation in the air flow less than down to 75% of the design rate. Thus two-speed fans would not be able to efficiently adjust to only ambient temperature driven variations. Even fan cycling would have limitations. Without fans operating, air flow through a tower will continue by the thermal draft at flow rates up to one fourth that of operating two-speed fans on low speed (14). Both would be sizeable reductions when only a maximum of 20 to 25% reduction in fan speed or around 50% reduction in fan power is typically required, as shown in Figures 5 and 6.

Figure 5 Seasonal Variation in Percentage of Required Fan Horsepower for Constant Heat Load (A)

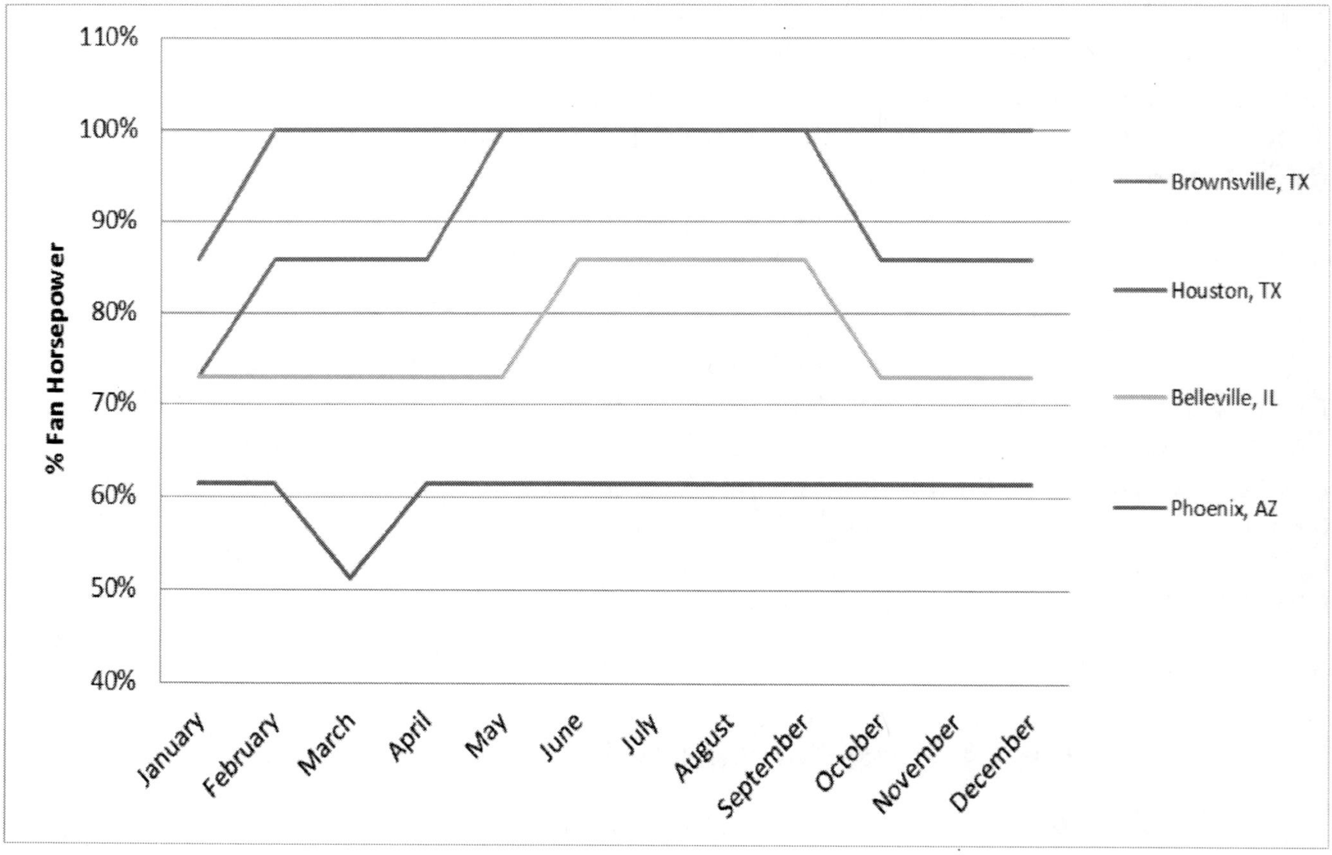

Figure 5 also summarizes the energy savings benefit of reducing tower capacity to match changing site ambient conditions. For these sites the energy demand is highest during summer conditions; with winter conditions being the season for lowest energy use, as well as the season for lowest water use. With the exception of the US Gulf Coast locations during summer, the monthly energy use usually averages more than 15% less than the equipment design capacity.

However, the sites graphed in Figure 6 either show constant energy use based on monthly average conditions or highest energy use during winter conditions.

Figure 6 Seasonal Variation in Percentage of Required Fan Horsepower for Constant Heat Load (B)

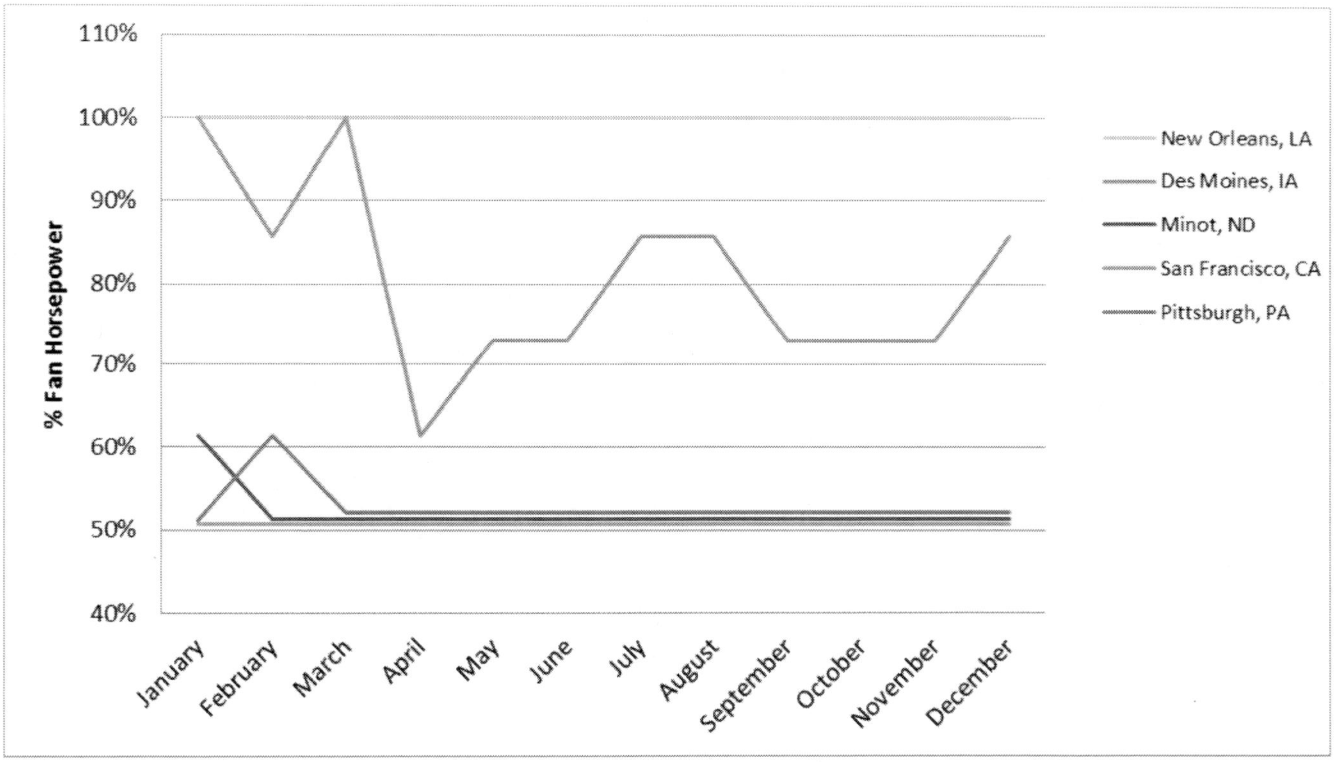

While New Orleans and San Francisco experience ambient weather conditions that trend to consistent energy use; sites such as Des Moines, Minot and Pittsburgh that experience cold and dry ambient weather trend a more complicated energy use graph. Although all the sites, except Minot, in Figures 5 and 6 experience maximum predicted fan energy use for the summer design condition; the average monthly predicted maximum energy use for Des Moines, Minot, and Pittsburgh might occur during the coldest winter months. This energy use pattern is attributed to the water absorption capacity of cold air decreasing quicker than the increasing ability of this cold air to remove heat by sensible heat transfer. More air flow would be required as the sensible heat transfer mechanism becomes dominant; even though less water is removed during these months. However, when the cooling water cold temperature falls into the low 40's F; many sites choose to save fan energy and reduce icing risk by allowing the cold water setpoint to rise. Most processes do not benefit from cooling water at these winter temperatures anyway (13). Therefore, Figure 6 likely overstates the predicted energy use for sites such as Des Moines, Minot and Pittsburgh during the winter months.

As shown in Table 5 for the same heat load, the predicted average fan energy required across the sites in July and January is similar; however, the percentage of water required to be evaporated in January averages 30% less than in July. Most of the sites require approximately the same fan energy in winter and summer, while two are predicted to require more in winter and two less.

Table 5 Comparison of predicted energy use vs. evaporative cooling water loss at nine US locations for the same cooling tower design

		Average July		**Average January**	
Location	**Summer Design (% HP)**	**% Evaporation**	**% HP**	**% Evaporation**	**% HP**
New Orleans, LA	100	1.3	100	1	100
Brownsville, TX	100	1.4	100	1.1	100
Houston, TX	100	1.4	100	1	85
Belleville, IL	100	1.3	100	0.8	75
Des Moines, IA	100	1.3	85	0.7	100
Phoenix, AZ	75	1.6	60	1.1	60
Pittsburgh, PA	60	1.2	50	0.8	50
Minot, ND	45	1.2	35	0.6	65
San Francisco, CA	35	1.1	30	0.9	30
Average	**80**	**1.3**	**75**	**0.9**	**75**

Climate

With cooling tower system performance dependent on local ambient weather conditions dealing with the weather is a daily business concern. However, a changing climate can take weather concerns to a different level especially when accompanied by growing regulatory, environmental, and supply pressures. A 2013 report (15a) highlighted that climate change is often perceived as a slow, gradual change in mean temperatures. However, a relatively small change in average temperatures results in large increases in the risk of extreme heat and substantial decreases in the risk of cold weather. The same report (15b) identified the Manufacturing and Industrials sector as the only major sector to expect both "increased operational costs" and "decreased production capacity" from a changing climate.

Climate forecasting can highlight long-term changes in such factors as temperature, humidity, or rainfall patterns to help quantify a changing risk profile. For assessing the sustainability of water supplies as part of a water management program; drought can raise the price of water and climate-related regulation can drive up the cost of energy (3).

Figure 7 Seasonal Variation of Percentage of Total Cooling Tower Heat Transfer Due to Latent Heat (Evaporative Cooling) for a Climate with a 2F Degree Rise in Ambient Wet-Bulb Temperature

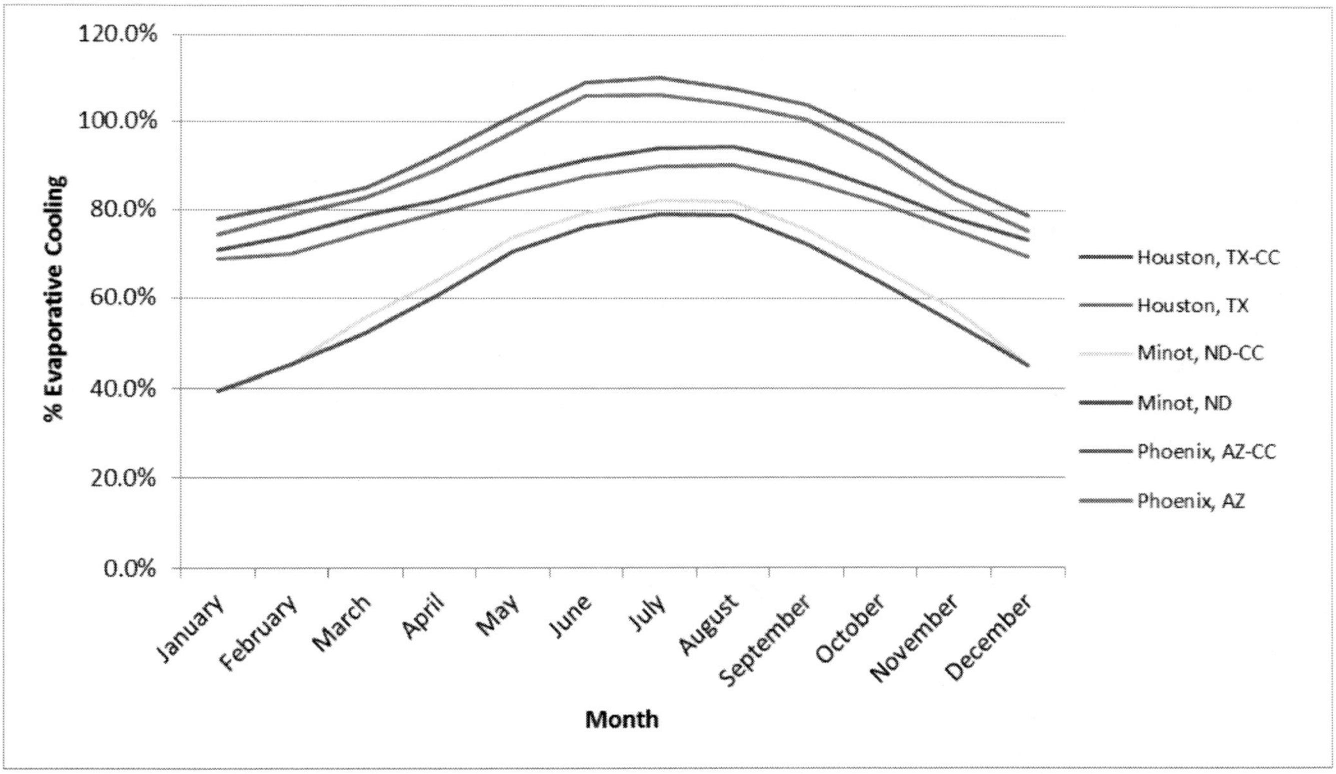

To project the impact of a warming, and for most regions moistening climate, over the next decades Figure 3 is modified as Figure 7 to show the projected increase in evaporative cooling required for a climate with a 2 F degree increase in ambient wet-bulb temperature. Though

more pronounced for the coldest part of the year, the predicted percentage increase in water use would average around 4% more than before the temperature change; as shown in Table 6. Assuming the same cooling tower equipment outlined in Table 1; the ability to increase the capacity of the cooling tower by this amount would rely entirely on more fan power use.

Table 6 Relative predicted evaporative cooling water loss with a 2F Degree Rise in Ambient Wet-Bulb Temperature at nine US locations for the same cooling tower design

		Relative Predicted Evaporative Water Loss (% Change)			
Location	**Summer Design (% Change)**	**Average July**	**Average January**	**Winter Design**	**Annual**
Des Moines, IA	4.1	3.7	9.3	0.3	4.4
Houston, TX	4.1	4.3	3.1	2.7	4.4
Brownsville, TX	4.1	4.2	3.6	6.6	4.6
New Orleans, LA	4.0	4.3	6.1	3.4	5.0
Belleville, IL	4.1	3.7	2.7	9.7	3.6
Minot, ND	3.4	4.0	0.7	14.5	4.0
Pittsburgh, PA	3.7	3.8	6.2	0.1	4.7
San Francisco, CA	3.9	4.5	5.2	5.9	4.8
Phoenix, AZ	3.8	3.7	4.6	2.4	3.7
Average	**3.9**	**4.0**	**4.6**	**5.1**	**4.3**

Concluding Thoughts

Based on a simplified model, the prediction of water evaporation loss in wet cooling towers as a function of ambient weather conditions has been discussed for constant heat load at ethylene production facilities. As shown in this paper, a cooling tower is designed to meet production cooling demands during extreme conditions and due to seasonal climate changes, the tower operates at design conditions infrequently leading to annual conservation of water and energy resources.

Historically, the primary factor considered in selecting cooling water systems was the availability of water and its associated cost. The analysis discussed here would enable the prediction of water and energy use of evaporative cooling tower systems to be available to assist in planning, siting, and permitting at the Front End Engineering Design level. As water and energy are being realized as essentially interdependent resources; considering monthly and annual use patterns provides an opportunity to approach the challenges of water resource management in a more integrated fashion.

References:

1. Gloyer, Walter (1978), Review of Cooling Tower Calculation, Cooling Tower Institute – TP78-06 (TP-194A), January.
2. Lakovic, L.S.; Mladen Stoiljkovic and Mirjana Lakovic (2004), Today's condition and the future of the cooling towers, Thermal Sciences 2004. Proceedings of the ASME - ZSIS International Thermal Science Seminar II, Bled, Slovenia, June 13-16, pages 701-704
3. Stone, John C. et al (1975), An Integrated Power Process Model for Ethylene Production, Water Resources Research Vol 11, No. 9 December.
4. ____ (2008), ASHRAE Handbook – HVAC Systems and Equipment, Chapter 39 – Cooling Towers.
5. Leeper, Stephen A. (1981), Wet Cooling Towers: Rule of Thumb Design and Simulation, US DOE EGG-GTH-5775 July
6. Engineering Weather Data Handbooks (2003) Air Force Combat Climatology Center (AFCCC) / National Climatic Data Center (NCDC). For each site:
    a. Section 2-2.2.1
    b. Figure 2-12. Sample Data Set Page 12
    c. Figure 2-1. Sample Data Set Page 1
7. ___ (1974) Development Document For Effluent Limitations Guidelines And New Source Performance Standards For The Major Organic Products Segment Of The Organic Chemicals Manufacturing Point Source Category, USEPA, April; Section III pg. 53.

8. Eftekharzadeh, Dr. Shahriar; Dr. Muin M. Baasiri; Paul Lindahl, Jr. (2003) Feasibility of Seawater Cooling Towers for Large-Scale Petrochemical Development, Paper TP03-17 Presented at the Cooling Technology Institute Annual Conference; San Antonio, Texas – February 10-13, 2003 p 11.

9. Jalal Engineering. Cooling Tower Basics and Common Misconceptions. oshuja@cyber.net.pk, pp. 1-6. Accessed online September 2016.

10. Cheremisinoff, Nicholas (2000), Handbook of Chemical Processing Equipment, Chapter 2 – Evaporative Cooling Equipment. Elsevier.

11. Schwedler, Mick (2014), Effect of Heat Rejection Load and Wet-Bulb on Cooling Tower Performance, ASHRAE Journal (January) pp. 16-22.

12. ASHRAE (2001), ASHRAE Handbook and Product Directory—Equipment, Chapter 21. American Society of Heating, Refrigerating and Air-Conditioning Engineers, Inc., USA

13. ___ (2013) Variable Flow Over Cooling Towers For Energy Savings, Publication TR-014 SPX Marley Cooling Technologies Inc., Overland Park, KS (March).

14. Midwest Cooling Towers Blog (2013), Cooling Tower Fundamentals: Cold Weather Operations and Components, January 22, http://midwesttowers.com/cooling-tower-fundamentals-cold-weather-operations-and-components/

15. Crawford, Meg and Stephen Seidel (July 2013), *Weathering the Storm: Building Business Resilience to Climate Change*, Center for Climate and Energy Solutions, c2es.org.
    a) Appendix B
    b) Figure 12

16. Engel, Hauke; Per-Anders Enkvist and Kimberly Henderson (July 2015), "How companies can adapt to climate change", McKinsey & Company.

17. Morrison, Frank (2014), Saving Energy with Cooling Towers, ASHRAE Journal (February) pp. 34-40.

AICHE PAPER NUMBER: 171d

OPTICAL GAS IMAGING: TAKING A DRIVE THROUGH IR CAMERA TECHNOLOGY AND HOW IT'S USED

Justin Lamb
Project Manager, GHG
Dexter ATC Field Services

Nick James
Operations Director
Dexter ATC Field Services

Prepared for Presentation at the 2018 Spring National Meeting
Orlando, Florida, April 25, 2018

OPTICAL GAS IMAGING: TAKING A DRIVE THROUGH IR CAMERA TECHNOLOGY AND HOW IT'S USED

Justin Lamb
Project Manager, GHG
Dexter ATC Field Services
Nick James
Operations Director
Dexter ATC Field Services

Abstract:

Have you heard about the latest Optical Gas Imaging (OGI) technology, but are still asking the question "What's the big deal?" Don't worry, you aren't alone! This presentation will drive you through federal and state OGI rules/regulations; exploring baseline and creative applications of this technology for industry use.

Along the way, we'll revisit the birth of infrared (IR) technology, how far we've come, and where OGI applications can take us in the future. We'll take a look at a few industry-leading brands of IR cameras; comparing operating capabilities and individual accessories/functions of each.

Real videos of recorded fugitive emissions will be presented to provide a visual representation of IR technology. Examples of leaks will be provided in multiple video layers to discuss the advantages and limitations of each filter applied. In addition, this discussion will address the concerns of OGI efficiency and/or effectiveness, as opposed to the traditional leak detection methods.

If you don't know where you're going, how can you get there? This presentation will map out the many different factors that can drive federal/state OGI monitoring requirements, as well as voluntary best practice programs. Topics such as safety considerations and product losses will be covered.

The road to OGI monitoring applications and programs should never be burdensome, but preparation is the key to success in this fast-moving industry. Join us as we go on this journey together because remember, protecting the environment is good business.

Optical Gas Imaging: Taking A Drive Through IR Camera Technology And How It's Used

Introduction

The goal of any industry, whether it be petro-chemical refining and production, storage and transfer, or exploration, is to optimize the overall process at the highest efficiency, with the lowest cost. Over the past few decades, world-leading engineers and scientists have developed new ways to produce and extract resources at rates previously not thought to be possible. As these advancements allow increased production rates and higher profits, the number of facilities designed to do so have grown exponentially. As most would agree, the ramped up production rates have caused public interest in the health of our environment. The U.S. Environmental Protection Agency (EPA) has responded to this interest by promulgating regulatory requirements to ensure these operations are limiting their effect on the atmosphere. Releases of volatile organic compounds (VOC) and other potentially harmful chemicals to workers, the public, and the atmosphere have been one of the leading factors that laid the foundation to Leak Detection and Repair (LDAR) requirements. As these requirements have been implemented, revised, and increased, the need for technology to provide us with the ability to do so has also grown.

Optical Gas Imaging (OGI) is a technology that has been designed help industry meet the demands of the EPA. The term OGI, coined by EPA in 2006, simply put is the act of making invisible gases, visible. This technology uses infrared (IR) waves to detect emissions or VOCs from leaking components on piping, vents on tanks, or combustion flares (to name a few) that are otherwise invisible to the naked eye. The IR camera is a non-contact device that detects infrared energy (heat) and converts into an electronic signal, which is then processed into a thermal image on a video monitor. To most, this cutting-edge technology has changed the way we see gases, almost as if we're viewing our surroundings in another dimension. OGI monitoring has traveled a long, not so narrow path from where its technology was discovered, to making it an industry-wide recognized way of performing and maintaining LDAR programs.

History Of The Camera

In order to understand how far this technology has come, it's important to know where it began. It in year 1800, Sir William Herschel of Germany, accidentally discovered IR light. Herschel, an astronomer by trade, wanted to know how much sunlight passed through different color filters on his telescope lens. He developed an experiment to record the temperature of light at each color on the visible light spectrum, and noted that as you move from the violet side of the spectrum all the way to the far red side, temperature recordings increased. He was curious to see what would happen if he went beyond the red spectrum, so he conducted an experiment he called "calorific rays", derived from the Latin word "heat". The results showed that beyond the red spectrum, there was radiation that displayed similar qualities (reflection, refraction, absorption, etc.) as that of the visible light spectrum. Herschel had officially discovered what today we call, infrared light or infrared radiation.

Fast forward 130 years, and we finally begin to see the application of this spectrum applied to an instrument, and used for a specific purpose. Following World War I, the first thermal imaging camera was produced by a Hungarian physicist for the British anti-aircraft defense. Also known as "night vision", the camera allowed the military to see thermal images that the rest of world didn't have access to. Following the 1930's, interest in the technology and possible advantages it could provide during war increased dramatically. In 1947 the United States and Texas Instruments developed the first ever infrared line-scanner, which was operated by a single element detector. It was soon realized that the primary limitation to the scanner was the time it took to produce a single image, which was roughly one hour at the time. As stays true for most inventions, once a limitation is evident, it's a figurative foot race to overcome and eliminate that road block. As new models were developed and the speed of image production increased, the use of thermal imaging cameras was applied to more practicable situations such as fighting fires or locating victims following a natural disaster. Even though applications for the technology were broadening, the military wanted sole possession. High prices and sensitive, possible dangerous applications were two primary reasons the military sought to keep it out of the public's hands. It wasn't until 1992 that the U.S. government officially declassified the technology, making it available for commercial sale and use. Shortly after the release to the public, Honeywell began licensing patents associated with microbolometers to several companies, which was originally developed in the late 1970's. While Honeywell focused on what was labeled vanadium oxide (VOx) microbolometer technology, Texas Instruments went after commercial production of infrared camera detectors. They were granted patents for detectors that were made of barium strontium titanate, or BST. BST was used to coat the thermal imager's sensor, which allowed for more absorption of light.

With the declassification of IR technology, following the Gulf War, demand for OGI cameras for commercial use increased dramatically. Commercialization forced production volume up, allowing for mass manufacturing, which in return minimized the cost per unit for both suppliers and consumers. As history tells it, this was a pivotal time in the life cycle of the camera's physical dimensions and practicality of use. Consistent with the evolution of most technology (personal computers, mobile phones, etc.), the desire to reduce the size and weight of the camera while maintaining core operating functionality saw a dramatic increase. Early equipment models were bulky, extremely heavy, and difficult to transport. The first industry-used cameras were so large in fact, that the only way to transport and ultimately use them was by stationary use. Some cameras were mounted to the top of vehicles (vans) and driven around facilities in order to perform OGI monitoring.

It's important to discuss what has been labeled as The Common Sense Initiative, an excerpt published by the U.S. EPA in April of 2006. As quoted by Carol M. Browner, an American lawyer, environmentalist, and businesswoman "We want to make good on what business and environmentalists have been telling us for two decades – that we must look at whole facilities, whole industries, and their overall impact on the environment. We must do a better job of cleaning up the environment and do it cheaper". Browner was referring to the burden of 25 plus years of environmental initiatives (regulations), the differences in cost required to comply for small businesses vs. large industry leaders, and the overall complexity and rigidness that came along with it. Such initiatives were founded on the intention of protecting our waterways, soils, and limiting air pollutants, but as history tells it, presented unintentional consequences. The goal of The Common Sense Initiative was to ensure a healthier environment, but without diminishing the health of the U.S. economy, which prior to this implementation, resulted in nearly 30 billion dollars spent for compliance by U.S. businesses. In order to comprehend the needs, limitations, and impact of this initiative, representatives from federal, state, environmental justice groups, labor force, industry, and community-based organizations teamed up to develop a plan that was all-inclusive and representative of nationwide operations. By developing a stream-line process to overall better the intentions of the agency, optical gas imaging popularity and discussions about its implementation were, for the first time, taken into serious consideration. Of the five major projects that the Common Sense Initiative (CSI) counsel undertook to resolve complexity and burdens, exploring alternative ways to meet regulatory compliance and standards gave OGI monitoring a solid footprint for future discussions. As we saw the momentum of these discussions increase by the CSI, it was now technology's turn to step up to the plate and make this alternative monitoring technique a practicable solution.

It wasn't until 2006 when FLIR Company (Forward Looking Infrared) introduced the benchmark OGI camera called the GasFindIR. This modern-looking, light-weight model was portable enough for users to carry in the field, while maintaining an all-inclusive battery system that, until then, wasn't feasible. It was the first model of its kind that was solely designed for the detection of VOCs inside refineries, chemical plants, gas processing facilities, transmission stations, storage facilities, and production wells. Following the release of the GasFindIR camera and the overwhelming feedback from users, the OGI manufacturing industry foot race was underway. This introduced opportunity for smaller start-ups to piggy-back off of existing technology, allowing for the development of similar instruments to be released to the market. Companies such as FLIR and OpGal (founded in 1983), who controlled the majority of the market, began to develop newer models that provided specific functions for all OGI needs. Cameras designed specifically for furnace and electrical inspections, refrigerant leak detection, continuous gas leak detection, and CO detection gave industry a wide variety of options to choose from, based on their specific needs. Recent years of advancement has given compatibility of cameras to smart phones, allowing real-time video of surveys to be transferred without the phone user to physically be there. Fast-forward to the present, and now we see various IR camera manufacturers and UAV drone companies working together to conduct surveys remotely. This development allows for operators to perform OGI monitoring in areas previously inaccessible by foot, which if you're familiar with the geographic locations of production well pads and pipelines in the U.S., is a major advantage. Not only did this transition increase productivity in scarce areas, it also provided an additional layer of safety for monitoring in areas known to be hazardous or immediately dangerous to life. As the old saying goes in the south "The proof is in the pudding". There has been years of research and development put into this technology to where we see it today, and if it wasn't a feasible, more efficient and safer practice, the evolution wouldn't have made it this far.

So what brings us to where we are now, and why does OGI monitoring tend to be in most major environmental discussions today? First and foremost, we've finally arrived at the "walking" stage in the life-cycle of this technology. As mentioned earlier, the sophistication of OGI technology has been steadily increasing over the past few decades. Numerous pilot studies such as the Common Sense Initiative, have proved that the use of infrared camera technology is capable of meeting the requirements of government agencies. Secondly, regulatory requirements are now finding themselves to be in parallel with the use of this monitoring technique. In the past, when OGI monitoring first made its debut in the Alternative Work Practice, it was merely an option to primary methods. The pressure that was put on EPA to release another method to maintain compliance was evident, and soon shown to be somewhat of a premature decision. While a new technique of doing things was underway, the requirements for reporting and record-keeping remained consistent with traditional practices. It was soon realized that though OGI monitoring was an acceptable practice, it only fulfilled a fraction of the various already-existing regulations. This meant that if operators wanted to use this surveying technique, they would still have to maintain their previous obligations of monitoring compliance, such as Method 21. Fairly enough, this discouraged a majority of companies to explore alternative methods, because of the simple fact that it didn't replace the system they were trying to escape from. Even though this presented a major roadblock in what was hoped to be a straight road, the use of OGI monitoring was still favorable in many situations. It wasn't until 2015, when EPA published the first ever, 100% OGI monitoring regulation, now commonly known as NSPS OOOOa. This requirement was aimed at reducing the amount of methane release into the atmosphere from the upstream and midstream oil & gas sectors nationwide. Probably the largest accomplishment in the eyes of manufacturers, who pushed their research for decades, to finally see a regulatory requirement for industry to use their product.

Rules And Regulations

It's no secret that the implementation of regulatory requirements comes at the expense of the businesses affected. It's reasonable to argue that by complying with these requirements, companies are losing profit. Costs associated with developing and maintaining environmental departments in the oil and gas industry, for example, have significantly affected the way these companies run their business. In the early days, there were simply no regulatory requirements to ensure extraction, transfer, and refining of crude oil was conducted in a manner that would limit contamination and/or pollution to the environment and surrounding public. Many could argue that the lack of environmental responsibility was due to limited research of chemical compounds handled, the public's elementary knowledge of the industry and how it worked, or by the bottlenecking and large donations for political gain. Whatever the reason(s), it's important to understand that most regulatory implications were birthed in a reactive manner. Reactive implications typically followed the release of an oil spill in our oceans, an explosion at a refinery, or long-term devastating contamination in our soils and rivers. Each time something like this happened, new regulations were put in place to limit the reoccurrence of similar events in the future. Consequently, every time a new regulation was born it added a layer of complexity for businesses to continue operations.

In regards to OGI monitoring regulations, we have seen a slower but just as complex life cycle. Industry has struggled to understand and comply with the multiple layers of regulatory sanctions imposed, but has played a pivotal role in this evolution. One simple way to break down how these requirements affect a business, is to break it down into agency levels. First and foremost, in regards to the U.S. EPA, we look at federal-wide regulations. These typically affect the lower 48 states in the same manner, no matter what. Now there are some exceptions for businesses that may have operations in multiple unique geographic locations. For example, the recent NSPS OOOOa federal regulation states that owners or operators are required to perform OGI surveys at compressor stations on a quarterly basis. The vast majority of businesses affected by this rule are capable of and required to do so every single calendar quarter. EPA does, however, provide an exemption from surveying compressor stations located in geographic regions with significantly lower average temperatures for a specific quarter. Reasons for this exemption may reflect inaccessible facilities with increased snow/ice on roads, and a drawback in actual compression operations, to name a few. After federal regulations, we see the next "level" down into state-wide regulatory requirements. States such as California (CARB-32), Colorado (Rule 7), and Pennsylvania (Exemption 38), have all led the way in developing their own set of requirements, in regards to OGI monitoring. In fact, all three of the states listed above were implemented before the federal NSPS OOOOa requirement, making it evident that a slow introduction was necessary before EPA felt they could gain national acceptance. Making our way down another level is where we can find tribal and local agencies implementing more specific rules. Typically, compliance requirements at this level are the result of public concerns and immediate need for some type of resolution. Although local (city, county, etc.) agencies don't have the level of authority that state and federal agencies do, we see organizations such as the Bureau of Land Management (BLM) retain a high level of respect. The Methane Rule, as named by BLM, mirrors that of NSPS OOOOa, but with a few different requirements for subjectivity. The compilation of multiple agency regulations has created pushback from industry, arguing that monitoring and reporting requirements are too complex, financially burdensome, and in many cases redundant. Businesses were finding themselves reporting the same information to multiple agencies, at multiple times of the year, with different record-keeping requirements. Even with the publication of NSPS OOOOa, industry is still required to follow the rules of whichever is most stringent, while maintaining compliance as stated by each agency level.

As industry has slowly adopted such regulatory requirements, an interesting and positive trend has come to light. While not all businesses and facilities have been subjected to maintaining OGI monitoring programs, a majority have adopted voluntary practices for proactive measures. Infrared technology has proved to be beneficial to businesses in ways not previously found. Today we see a majority of industry voluntarily investing in the technology to better their business, seeing positive results in both the short and long term aspects. Research has shown that implementing the use of OGI programs has consequently improved the financial health of businesses that opt into this practice. Conoco Phillips Company and FLIR teamed up to conduct a pilot study to determine the cost benefit analysis of implementing such a program, and found that the return on investment for OGI was extremely beneficial. This research illustrated that "The study identified 144 leaking components. Collectively, these leaks account 58.26 mmcf/y and $358,012.10 USD/year in lost product. The methane leak sources contribute 21,420.7 tonnes/year CO2e to GHG emissions. It is estimated that 92% of the 144 fugutive sources are economical to repair. Implementing all economical repairs would result in a net present savings of $2,002,602.72 USD". Another study, conducted by Jonah Energy and FLIR showed "The cumulative gas savings realized by the program has exceeded $5 million in the past 6 years, which has more than covered the overall program costs. This includes the Optical Gas Imaging equipment and associated operators, along with all repairs and maintenance, including labor and parts." Although these are only two examples of the benefits for voluntary implementation, the numbers speak for themselves.

Aside from limiting product loss and increasing revenue, the majority of businesses strive to do so in a safe manner. In recent years, we've seen worker safety pushed to the forefront of day-to-day operations, ensuring all employees return home safely to their families. Not only does implementing safer work practices benefit the health of those in immediate contact of hazardous processes, it also aids in ensuring the company is positively viewed in the public's eye. Utilizing OGI monitoring has allowed for surveying at safer distances, which isn't true for traditional M21 monitoring. It serves as an important tool during facility-wide turnarounds when pressuring up, allowing operators to quickly and safely detect possible releases from malfunctioning process equipment and loosely tightened connectors. Another example of how OGI monitoring has provided safety measures for both industry and surrounding communities is that of the infamous Aliso Canyon Blowout, near Los Angeles, California. During a routine flyover, scientists conducting OGI surveys on pipeline came across one of the largest known methane leaks known today. Also known as the Porter Ranch gas blowout, Southern California Gas Company's well was discovered leaking on October 23, 2015. During a 2018 earnings call, Sempra Energy (parent company) reported that costs associated to this catastrophe has reached nearly $1 Billion dollars. Lost revenue, costs associated to the remediation efforts, restitution, as well as penalties and fines all account for that figure. The leak was so large, it is estimated that more than 100,000 metric tons of methane were released into the atmosphere, requiring immediate evacuation from nearby communities in the San Fernando Valley.

OGI Monitoring vs Method 21

As industry either fought for or fought against the implementation of OGI monitoring, the technology's effectiveness for leak surveying was of high priority. The only feasible way to build an argument is to compare OGI monitoring to traditional and existing methods used in industry today. The technique used, for comparative purposes, is that of Method 21. Method 21 is a widely used term in the LDAR industry that describes the approved monitoring protocol on equipment (valves, connectors, pumps, PRVs, open-ended lines, etc.) that have the potential for emitting VOCs or HAPs (Hazardous Air Pollutants) into the atmosphere. One example of an instrument designed for Method 21 is that of the Thermal Toxic Analyzer 1000B, or TVA-1000B. Simply put, it's a machine that measures the concentration of a leak in parts per million (ppm) of any given leak (VOC/HAP) on a component. In general, facilities that are subjected to using this method, are required to perform inspections on all equipment that fall into this category, dependent on the stream composition of each. Since Method 21 is the primary option for LDAR compliance monitoring programs, it makes sense that all arguments be centered around it for OGI monitoring discussions. A simple yet effective way of comparing the two, was for EPA and industry to list the Pros and Cons of each. As previously suggested, OGI monitoring allows the user to visibly see the leak, the source of it, and the direction it comes from. Since "seeing is believing", this method allows for little to zero doubt that emissions exist, and provides real-time accuracy of the origination. While nobody is discrediting the accuracy of Method 21 monitoring, it's evident that it comes with its limitations. Compliance programs designed for Method 21 are required to have a complete inventory of all equipment subjected to monitoring, identified with a metal tag, and maintained in a robust database. After this phase is completed, technicians are required to monitor the components on a scheduled basis, depending on the regulation they fall into. The TVA-1000B for example, is designed to provide readings in ppm only, not always allowing the user to determine the exact point or origin of the leak. Furthermore, Method 21 is limited to locating fugitive emissions from only the tagged and documented components in the compliance database. Components that don't fall into the program still have the opportunity to emit harmful toxins, but there's no way to identify the geographic location of these if they aren't being regularly monitored. When performing Method 21, the user can sometimes pick up high readings near a tagged component that isn't actually leaking, which urges the question "Where is this coming from?" Utilizing OGI methods allows users to perform surveys on all components, not just the ones regulated in the program. This provides the capability to detect and resolve leaks that could be problematic for years, but otherwise wouldn't be found if only using traditional methods.

A major setback for using the camera to monitor is the reality of not being able to measure the concentration of the leak, as the TVA-1000B is designed to do. This presents a gap in the argument of what constitutes a leak, and what is an acceptable limit. In a controlled environment, the OGI camera is capable of detecting leaks at or around 1500 ppm. In personal experience using the camera for monitoring, most leaks aren't found unless they're at least 5000 ppm or higher in real-time weather conditions. Not only is the majority of Method 21 monitoring conducted in an uncontrolled environment, the benchmark leak limit nation-wide is set at 500 ppm. Obviously, the gap in detection limits between the two methods presents a hesitation for allowing either practice. Environmental conditions play a huge role in how effective and accurate an IR camera can be throughout a facility. The wind speed, wind direction, temperature, barometric pressure, and humidity all play a factor in how well the user can see emissions through the camera lens. In an effort to limit this concern, the EPA publishes the Alternative Work Practice (AWP) for the oil & gas sectors in 2006. In the AWP, if a business opts to use

OGI monitoring, they're required to perform a daily camera check before monitoring commences. Unlike the TVA-1000B, which requires morning and afternoon machine calibrations with methane gases, this camera check is designed to ensure the user is monitoring in a manner directly related to real-time weather conditions. For example, a calibration gas (concentration 99% methane) is set up in the field with a flow meter attached before camera surveying commences. The user enters in the current environmental data, then sets the flow meter to a specific range to represent real-time conditions. The user then stands back 15-30 ft from the calibration setup to ensure the methane plume is visible from a specific distance; doing so in the mode they intend on performing the survey. This task allows the user to determine their survey distance from components in the actual conditions they're monitoring in, ensuring they don't exceed the distance from what the camera is capable of "seeing". So unlike calibrating the TVA-1000B to specific gas ranges, performing the daily camera check is really just telling the user that they can monitor from a specific distance in specific conditions.

Just as the debate on whether or not OGI monitoring is an effective way to perform surveys, the conversation on how efficient this practice is followed. Before looking at ways to compare efficiency, it's important to distinguish the two types of efficiency discussed; monitoring/use efficiency and program efficiency. The monitoring efficiency of OGI practices in many ways, trumps that of Method 21 monitoring. Studies show that the average number of components monitored/day using Method 21 is anywhere from 200 to 400 components/day, with OGI monitoring being ~15,000 components/day. With Method 21 a user is required to physically walk up to each components to collect readings, while a technician using OGI can view each component from a distance of ~30 feet away (note: assuming leaks are medium in size). Another pro of using the camera vs. Method 21 has been shown during unit or facility-wide shutdowns and turnarounds. Something as intensive and high-risk as replacing major process equipment can present potentially catastrophic events if taken lightly. One important risk that OGI monitoring can help reduce is that of pressure-up tests, when the construction is completed and the unit is ready to be tested. During such pressure tests, performing OGI-monitoring surveys on newly added vessels and associated equipment can ensure everything is working as designed. The chance that not all equipment has been tightened down securely following construction is always there, especially with many contractors performing the work. Miscommunication between operations and contractors is common, so the ability to utilize OGI technology to ensure proper maintenance is conducted can be a huge benefit during these times.

For purposes of this paper, program efficiency is referred to the techniques, training and overall continuation of maintaining an OGI program. Facilities required to maintain an LDAR program via Method 21, know that monitoring technicians are put through official Method 21 training. This training educates them on the proper technique for surveying each component, the speed of how to do so, and the time required to spend on each. As the majority of technicians performing these inspections are entry-level candidates, usually not required to have any sort of technical degree, the expectations for process knowledge and chemicals within facilities are limited. Although basic chemical awareness is required, technicians aren't expected to understand the advanced engineering methods and chemical mixtures of each process lines in which they're monitoring. LDAR software now has the capability to direct technicians to each component, provides them with a time lapse of monitoring requirements, and automatically logs leaks if they exceed the defined emission limit. With OGI monitoring, the training expectations are much more intensive than that of Method 21. Since not all cameras are created equal, it's important for the technicians to possess knowledge of stream speciation and how the overall process is engineered. All chemicals have what is called a Minimum Detected Leak Rate (MDLR) in accordance with the camera. In order to perform OGI surveys in a satisfactory manner, the technicians should know exactly what's in the streams they're monitoring. Likewise, it's also very important to understand the chemicals you're working with because there isn't one camera that can detect everything. Some instruments are made specifically for electrical and thermal inspections, some are designed for Carbon Monoxide, CO2 or Ammonia only. The need for initial training and education is higher and more intensive for OGI monitoring, which could pose higher administrative costs.

Recordkeeping And Reporting

There's an old saying in the LDAR industry that suggests "You're only as good as your documentation". By this we mean, it doesn't matter what you do with your compliance program unless you can prove it. Over the years, software databases geared to house large LDAR programs utilizing Method 21 have evolved in a manner which takes a majority of the risk away from businesses. The databases work directly with the field software used, and compile monitoring data in an easy-to-read format for review. Although somewhat sophisticated programs, traditional LDAR databases are limited to traditional monitoring methods. With the introduction of OGI monitoring techniques, we're now presented with the task of managing our data. Such record-keeping requirements can include: monitoring dates, equipment used, calibration records, list of components monitored, leaks detected, dates of repair, GPS coordinates of leaks, pictures of leaks, videos of leaks, videos of repaired leaks, videos of entire surveys, deviation from monitoring plans, etc. Until recently, there weren't any LDAR databases or software capable of storing these records. Most businesses had to create large files to retain hours of video footage, hand drawn maps, and custom leak documentation. Not only was this burdensome to hold large files of data, but it was unorganized and hard to report if requested. With the introduction of new OGI requirements and the overall sense of its long-term stay, new databases have emerged that have the capability of managing this data. Field tablets with GPS capability linked to GoogleEarth can now map out survey routes, with time stamps and leak footage. The capability of compiling this data into the expected reporting format is now available with the click of a button, which has been a main point of concern for businesses. The argument that it's impossible to streamline OGI monitoring data to meet reporting expectations no longer exists, and we're still only in the beginning stages.

Choosing Your Equipment

Before making the decision to purchase your first IR camera, there's homework to be done first. As mentioned earlier, not at cameras are created equally. As the IR manufacturing industry began to lift off in the early 90's, it was merely a race to the consumer. The goal was to provide the smallest, lightest camera, capable of monitoring in the infrared mode before the next guy. Now that we've come to understand the ability of the technology and made it a reality, we see a rise in expectations of using it. Two of the industry's leading manufacturers, as mentioned previously, are the FLIR and OpGal companies. In 2006 when FLIR introduced the GasFindIR model, it was designed specifically for the detection of VOCs in the oil & gas industry. As luck would have it, not all industries are created equal either. Different chemical compositions, processes, and engineering are seen throughout the world, facility to facility. Implementing OGI usage from a crude oil refinery to an Ethylene plant introduces the possibility of needing specific camera models, dependent on gases to be monitored. While the FLIR GF-320 and Opgal EyeCGas models are the most sophisticated and popular, they aren't a one-stop shop for meeting your goals. It's extremely important to know the exact chemicals you wish to detect, and then steer your shopping in that specific direction. I encourage you to visit the manufacturer specifications of each model, studying the limitations and advantages of each. Not only will this ensure you have the right equipment to fulfill the desired goal, but it may save on costs associated with each. If the determination is to survey for a single chemical composition or a specific task, you may find a particular model cheaper that's more affordable while still accomplishing the task. As we continue to make stride towards this "new age" technology and ways to implement into every day operations, there's little doubt that our workplace can be more efficient, effective, safer, and profitable.

References:

1. United States Environmental Protection Agency (October, 2007), *Leak Detection and Repair: A Best Practices Guide,* Office of Compliance, EPA-305-D-07-001
2. Watton, R., Smith, C., Jones, G.R., (1974, Dec. 9-11). *The pyroelectric vidicon: Pyroelctric materials for operation at cathode potential in a "Hard" tube.* Retrieved from https://ieeexplore.ieee.org/document/1477887/.
3. Know Made Patent & Technology Intelligence. (2015, July). Honey Microbolometer Patent Portfolio Analysis [Web Log Post]. Retrieved April 17, 2018, from http://www.knowmade.com/downloads/honeywell-microbolometer-patent-portfolio-analysis/
4. National Service Center for Environmental Publications. (April, 1996), *The Common Sense Initiative: A New Generation of Environmental Protection.* Retrieved from https://nepis.epa.gov
5. Grigoryants, O. (2018, February 28). *Here's the bill so far for the Aliso Canyon gas leak: Nearly $1 billion – and counting.* Retrieved from https://www.dailynews .com
6. Hazburn, C. (2018, February 19). *The History of Thermal Imaging Cameras.* Retrieved from https://www.ecamsecure.com/blog/history-thermal-imaging-cameras/
7. U.S. EPA, Code of Federal Regulation, Title 40, Part 60, Appendix A, Method 21 "Determination of Volatile Organic Compound Leaks". 2014 Ed.
8. McKenna, J., Gromicko, N. *The History of Infrared Thermography.* Retrieved April 1, 2018, from https://www.nachi.org/history-ir.htm
9. Palmer Wahl Instrumentation Group. (May 2017). *The Evolution of Thermal Imaging Cameras.* Retrieved from http://instrumentation.com/PDFS/ Evolution ThermalImagingCameras.pdf
10. Trefiak, T. (2015, December 14). *LDAR Case Study Comparison of Conventional Method 21 vs Alternative Work Practice (Optical Gas Imaging).* Retrieved from https://www.epa.gov/sites/production/files/2016-04/documents/20trefiak.pdf
11. O'Neill, G.R., Natural Gas STAR Methane Challenge Program. (2017, October 26). *Optical Gas Imaging History, Innovations and Industry Results.* Retrieved from https://www.epa.gov/sites/production/files/201711/documents/17.oneil_2017aiw.pd

514115 Implementing a Benzene Fenceline Passive Sampling Program

Abstract:

EPA has implemented a rule that requires refineries to conduct continuous passive monitoring of benzene at multiple fenceline locations using a new passive sorbent based Method (EPA Method 325) in order to control fugitive emissions that migrate past a facility border. Although, the rule is currently only applicable to petroleum refineries, this approach to monitoring may be extended to other facilities in the future. In this presentation we will touch on passive sampling theory, provide a regulatory breakdown. We will also discuss the current rule strategy or requirements (EPA Method 325a) for number and placement of samplers using specific site examples. We will discuss pre regulation information gathering or pilot study strategies as well as some discussion of corrective action if a â€œHOTâ€ spot or High benzene levels are found.

511492 Combined Experimental and Theoretical Thermophysical Properties Investigation of Deep Eutectic Solvents for Gas Capture Applications

Abstract:

Deep eutectic solvent (DES) produced by using various hydrogen bond acceptors (HBA) (e.g. betaine, alanine, choline chloride) with organic acids and sugars (e.g. citric acid, phenylacetic acid, levulinic acid and fructose) at various molar mixing ratios. Thermophysical properties including water content, thermal stability, density and gas solubility of CO_2 and N_2 were experimented at different isotherms for wide pressures range up to 50 bars. Moreover, detailed rheological experiments were conducted on the studied materials to obtain viscosity. Moreover, a density functional theory (DFT) and classical molecular dynamics (MD) methods have been used for investigating the physicochemical properties, structuring, dynamics and interfacial behavior of the studied NADES from the nanoscopic point of view to infer its viability for extensive usage. A pressure driven physisorption mechanism was observed and for the studied eutectic solvents, which were repeatable for various times without any degradation in the solubility performance. Experimental studies showed that presented materials appreciable CO_2 capture performance at low corrosion effect when compared with the most common amine-based CO_2 capture agents. No hysteresis was observed and no significant swelling tendency of the samples was recorded. Rheological experiments were studied to describe the dynamic flow behavior of the solvents. CO_2 absorption mechanism is evaluated by studying the behavior of the liquid â€“ gas and interface. DFT simulations yielded with an accurate quantification of short-range interaction; liquid state was also characterized via middle-range as well as long-range interactions together with volumetric effects. The low cost, non-toxic and low-corrosive physical properties of the presented solvents can be considered as an environmentally promising alternative for effective CO_2 capture agents.

512452 Synthesis, Characterization and Evaluation of (Ce, Y, Zn)-Ion-Exchanged Silicalite-1 Supported Ni Catalyst Structures for CO_2 Capturing By Methane Dry Reforming

Abstract:

Recent developments in nano-catalysis helped to re-design catalytic structures that can overcome conventional catalyst deactivation challenges, and hence, give rise to industrial reactions that are more environmentally friendly. The methane dry reforming reaction (MDR) is a useful way to produce H_2 and synthesis gas, and at the same time capture two main greenhouse gases; CO_2 and CH_4. However, through the last decades the reaction was not feasible due the rapid deactivation of the Ni based catalyst by fouling and sintering. In this study a microwave-assisted hydrothermal synthesis method was used to synthesize a Ni based catalyst, in which Ni at different loadings (10, 15 and 20 wt%) was deposited by incipient wet-impregnation on a MFI type silicalite-1 zeolite support. The catalyst performance was further improved by the introduction of metals; Ce, Y and Zn through an ion-exchange procedure. The alkalinity of the support was further enhanced by the introduction of Mg and Ca. These catalyst systems were evaluated at different temperatures and feed ratios in a packed bed reactor experimental setup. Different catalyst structural properties were characterized by XRD, FESEM, TEM, N_2-physisorption, XPS, FTIR and H_2-TPR. Time-on-stream evaluation experiments showed that the Ce ion-exchanged silicalite-1 supported Ni catalyst had the highest activity and stability. The catalyst maintained a CH_4 conversion of 90%, a CO_2 conversion of 85% and a H_2/CO ratio that ranged from 1.7 to 2.9. Ce improved the oxygen storage capacity of the catalyst which facilitated the oxidation of carbon depositions. In addition, the unique MFI structure that is composed of 10 straight membered channels of silicate-rings and 10 zigzag sinusoidal membered ring channels reduced diffusion limitations and offered a fast gas flow reaction environment. Carbon depositions from CH_4 cracking were also reduced due to the absence of alumina in the silicalite-1 support framework. Moreover, the high internal surface area of the support facilitated high Ni dispersion which prevented sintering of the catalyst. Raman spectroscopy, TGA and TEM images revealed that carbon deposits on most spent catalyst structures were predominantly multi-walled carbon nanotubes (MWCNTs) with defects.

Effects of Fossil Fuel Emissions and Seawater Absorption on Atmospheric Carbon Dioxide Concentrations

Dr. Frank J. Elvin, Texas Specialty Catalysts, Bellaire, TX

Abstract

A differential equation $\delta CO_2 = \delta Ef + 5.1 - 0.018CO_2$ represents the increase in atmospheric carbon dioxide concentrations in any selected year between 1750 and 2017. In this equation δCO_2 is the increase in atmospheric carbon dioxide concentration, δEf is the fossil fuel CO_2 emissions in ppm, and CO_2 is the atmospheric carbon dioxide concentration in January of the selected year. Although the equation was derived mathematically from published CO_2 data, the constants in the equation were verified by CO_2 seawater absorption and desorption experiments. Samples from the Pacific, Atlantic, Gulf of Mexico, North Sea, Baltic Sea, Black Sea and the St. Lawrence Seaway were tested in-situ and under laboratory conditions. The derivation of the equation, its experimental verification, and the relevance of CO_2 absorption by seawater to absorption by land-based biomass are discussed in this paper. The high degree of accuracy of the derived equation and the results of the CO_2 absorption experiments are consistent with the following conclusions. The CO_2 removal rate was a function of atmospheric CO_2 concentration, and neither the removal rate constant nor the effective absorption area changed between 1800 and 2015. The total annual rate of non-fossil fuel emissions not controlled by desorption kinetics, such as volcanic CO_2, was also constant. The annual fossil fuel CO_2 emissions rate was the only independent variable affecting atmospheric CO_2 concentrations, suggesting fossil fuel CO_2 was the primary cause of atmospheric CO_2 increases. Estimates of future atmospheric CO_2 concentrations were made for different fossil fuel emission rates assuming the equation remains valid. For example, atmospheric CO_2 could be reduced to 400ppm and maintained at 400ppm by reducing fossil fuel CO_2 emissions to 15Gt/year from approximately 35Gt/year. The longer the reduction in fossil fuel emissions is delayed, the higher atmospheric CO_2 concentration will rise.

Introduction

In 2011 the following equation was developed to represent the annual change in atmospheric CO_2 concentration. $\delta CO_2 = \delta Ef + K_1 - KCO_2$, where δCO_2 is the increase in atmospheric CO_2 concentration in ppm in a calendar year, δEf is the fossil fuel CO_2 emissions in ppm in that year, and CO_2 is the average atmospheric carbon dioxide concentration in January of that year. This generalized equation is based on mass transfer theory and reaction kinetics described by Levenspiel (1), Lewis and Whitman (2) and Fick (3). K_1 represents non-fossil fuel emissions into the atmosphere and KCO_2 represents the removal of CO_2 from the atmosphere. The two constants K_1 and K were calculated from published values of δCO_2, δEf and CO_2 for the time period 1800 to 1980 by solving sets of simultaneous CO_2 mass balance equations. Prior to 1958, atmospheric CO_2 concentrations were from ice-core CO_2 measurements (4) and post 1958 atmospheric CO_2 concentrations were measured at Mauna Loa (5). Annual fossil fuel emissions δEf were from Energy Information Administration (EIA) (6), Carbon Dioxide Information Analysis Center (CDIAC) (7) and International Energy Agency (IEA) (8). The annual CO_2 mass balance equation for the 1800 to 1980 data was $\delta CO_2 = \delta Ef + 5.0 - 0.017CO_2$, which is designated the 1980 equation. This equation suggests that the annual fossil fuel CO_2 emissions rate was the only independent variable affecting atmospheric CO_2 concentrations between 1800 and 1980.. A conversion rate of 7.799Gt of CO_2 for 1ppm of CO_2 in the atmosphere is used throughout this paper. One Gt is 10^9 metric tons.

In 2012 atmospheric CO_2 concentrations were predicted for every January from 1980 to 2011 using 31 consecutive stepwise calculations of $(CO_2)_{n+1} = (CO_2)_n + (\delta CO_2)_n$, where $(\delta CO_2)_n = \delta Ef_n + 5.0 - 0.017(CO_2)_n$. The subscript n represents the year. The 1980 atmospheric CO_2 concentration of 337.8ppm was the starting point for the sequential calculations, and was the only value of atmospheric CO_2 concentration used as input for this prediction. The other input data used was the annual published fossil fuel CO_2 emissions for 1980 to 2010. The predicted 2011 atmospheric CO_2 was 392.3ppm compared to a measured value of 391.3ppm as shown in figure 1. An accurate prediction was also made for 2011, starting with the 1750 value of 276.8ppm, which is shown in figure 2.

In 2017 a second equation was derived. This paper describes the 2017 study to determine if the equation or equation constants changed between 1750 and 2016, or are likely to change after 2017. The effects of seawater absorption on atmospheric CO_2 concentrations are analyzed and the absorption and desorption rate constants are calculated. Atmospheric CO_2 concentrations and global mean surface temperatures are predicted through 2100 for different CO_2 emissions rates. The relationship between atmospheric CO_2 concentration, the oceans' CO_2 vapor pressures and pHs are established.

Equation derivations, equation solutions, data, calculations, statistics and sensitivity analyses are included in supplementary Excel and word files. These files are not attached to this paper but may be available on request.

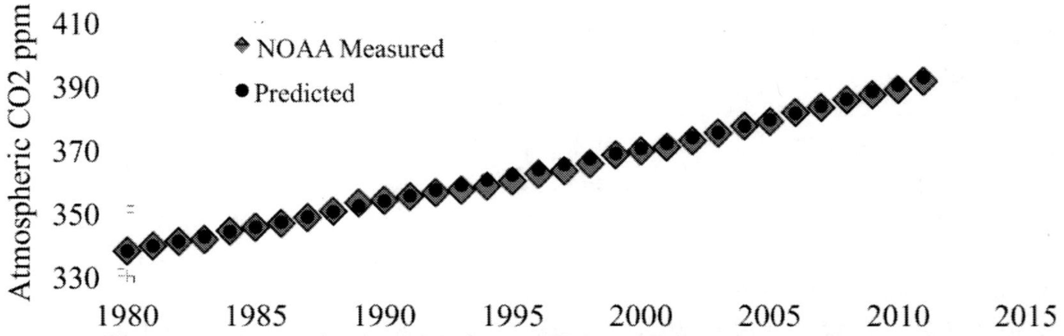

Figure 1 Atmospheric CO2 from 1980 to 2011 using 1980 equation

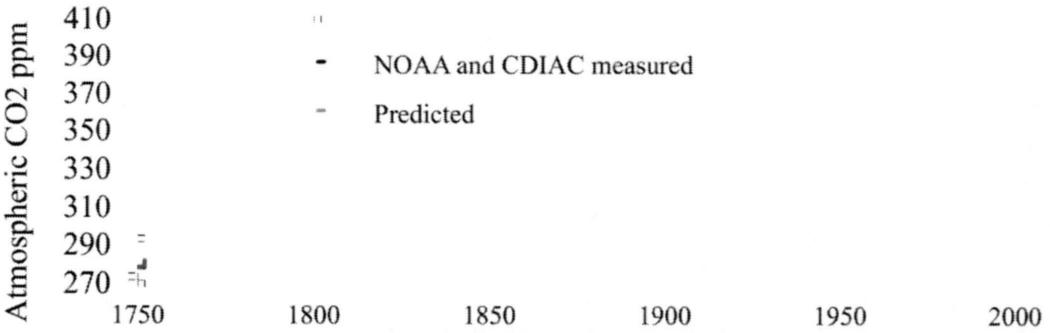

Figure 2 Atmospheric CO2 from 1750 to 2011 using 1980 equation

Prediction of the January 2017 atmospheric CO_2 concentration using the 1980 equation

Because of the excellent predictions from 1750 to 2011, the 1980 equation was used to predict the atmospheric CO_2 concentration in January of every calendar year from 1750 to 2017. The CO_2 concentration in the year 1751 is the value for 1750 plus the published fossil fuel emissions rate in 1750 plus 5.0 minus 0.017 times the January 1750 atmospheric CO_2 concentration. All values of January CO_2 concentrations can be predicted from the previous values. Therefore, it is possible to predict the 2017 value of CO_2 by performing 267 consecutive stepwise calculations from the published 1750 value of 276.8ppm without using any interim published values of CO_2. A plot of the predicted and measured values of atmospheric CO_2 is shown in figure 3. The predicted atmospheric CO_2 for 2017 was 407.6 compared to an actual measured value of 406.1ppm. In fact all 267 predicted values between 1751 and 2017 correlated to the published measured values with a $0.995R^2$. The standard error was 3ppm, which was less than 1% of the measured value. The mean deviation from the measured data was 0.0 and the mean absolute deviation was 5.8ppm. The high degree of accuracy exhibited by the 1980 equation between 1750 and 2011 continued through 2017.

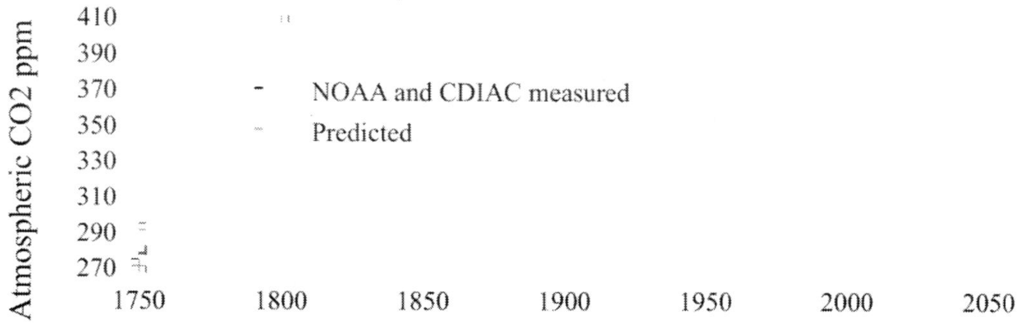

Figure 3 Atmospheric CO2 1750 to 2017 using 1980 equation

Derivation of the 2017 CO_2 equation using data regression

Even though the 1980 equation had predicted accurately through 2017, a second equation was derived using 1750 to 2017 data. Rather than solving simultaneous equations, as in the 1980 case, all the published data were regressed. To achieve a linear relationship, the CO_2 mass balance equation was transformed from $\delta CO_2 = \delta Ef + K_1 - KCO_2$ to $(\delta CO_2 - \delta Ef) = K_1 - KCO_2$. The constants K_1 and K were then derived by simple linear regression analysis. This is shown in figure 4 for five year smoothed $(\delta CO_2 - \delta Ef)$ data. δCO_2 = annual CO_2 increase in ppm, and δEf is the annual fossil fuel emissions in ppm. The result of the regression is $(\delta CO_2 - \delta Ef) = 5.13 - 0.018CO_2$, or $\delta CO_2 = \delta Ef + 5.13 - 0.018CO_2$. This is designated the 2017 equation. The 2017 equation appears similar to the 1980 equation. However, to confirm this predictions were made from 1750 through 2100 using both equations.

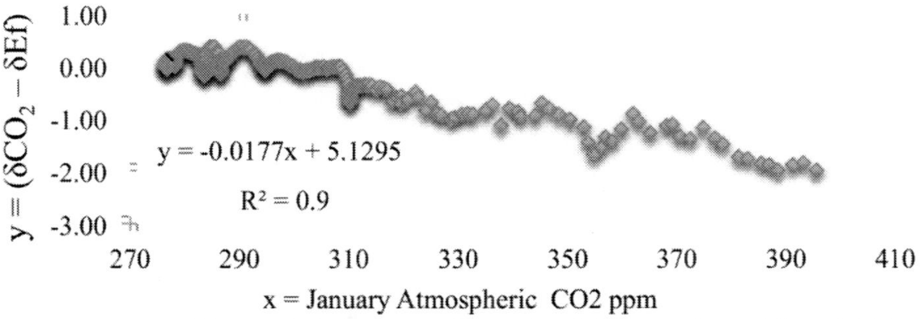

Figure 4 Five year Smoothed Regression 1750 - 2017

Atmospheric CO_2 predictions from 2017 to 2100

The difference between the two equations is insignificant even for predictions through 2100. This is shown in figures 5 and 6. Figure 5 predicts the atmospheric CO_2 concentrations from 2017 to 2100 assuming fossil fuel CO_2 emissions increase at 1.8%/year (the pre 2011 average increase). Figure 6 predicts the atmospheric CO_2 assuming fossil fuel CO_2 emissions do not increase and remain constant.

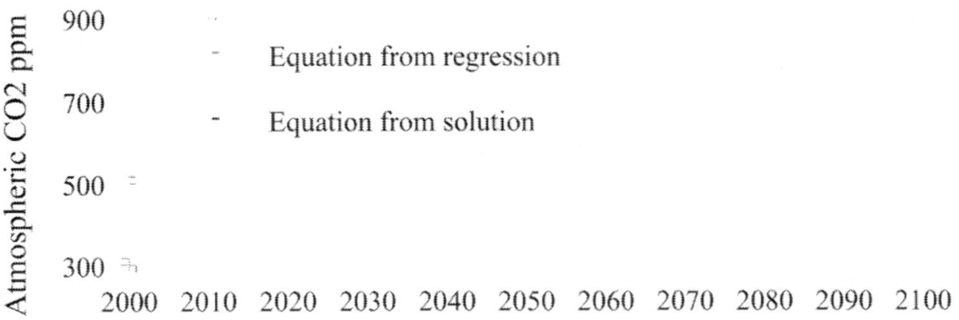

Figure 5 Atmospheric CO2 concentrations: 2000 to 2100 assuming 1.8%/year increase in fossil fuel CO2 emission

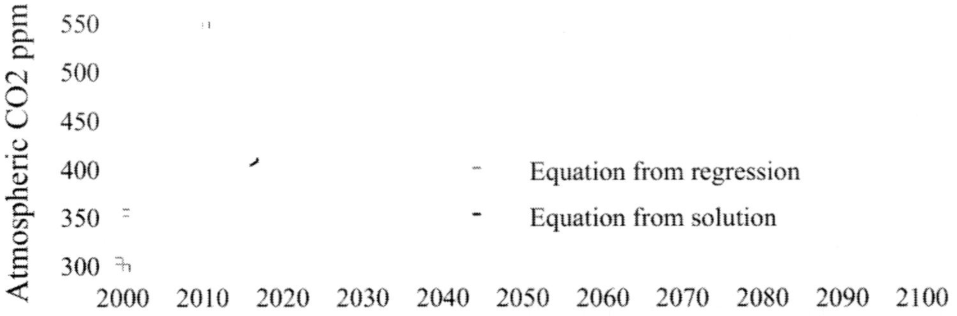

Figure 6 Atmospheric CO2 concentrations: 2000 to 2100 assuming no change in fossil fuel CO2 emissions

Atmospheric CO_2 predictions from 2017 to 2100 for different fossil fuel CO_2 emissions rates

From 2013 to 2015, global fossil fuel CO_2 emissions were nearly constant at 33Gt/year although they increased to 35Gt/year in 2016. Figure 7 shows predicted atmospheric CO_2 for several scenarios. If fossil fuel emissions remain at 33Gt/year from 2018 through 2100, and there are no changes in other factors, atmospheric CO_2 will be 501ppm in 2100. Atmospheric CO_2 will ultimately reach equilibrium at 529ppm. However, if annual fossil fuel emissions were to increase at 1.8%/year, which was the average annual increase in fossil fuel emissions between 1980 and 2010, atmospheric CO_2 would be 838ppm by 2100 and continue to increase. A 54% reduction in fossil fuel emissions from 33Gt/year to 15.16Gt/year in 2018, and maintaining 15.16Gt/year indefinitely, atmospheric CO_2 will equilibrate at 400ppm. If fossil fuel CO_2 emissions were eliminated in 2018, CO_2 would be 315ppm in 2100 and would equilibrate at 290ppm.

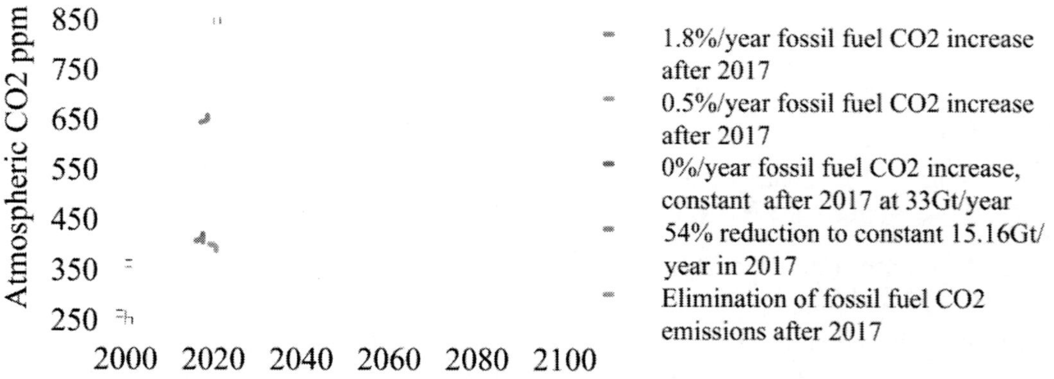

Figure 7 Atmospheric CO2 concentrations: 2000 to 2100

Figure 8 shows how atmospheric CO_2 would change if fossil fuel CO_2 emissions remain constant at 33Gt/year until they are reduced to 15.16Gt/year in one of three years: 2018; 2050; or 2100. In these predictions it is assumed the emissions will remain at 15.16Gt/year after they have been reduced. In all cases atmospheric CO_2 would equilibrate at 400ppm. However, because the rate of CO_2 removal is relatively slow at 1,7% or 1.8% of the atmospheric CO_2 content per year, it will take from 150 to 200 years to achieve equilibrium.

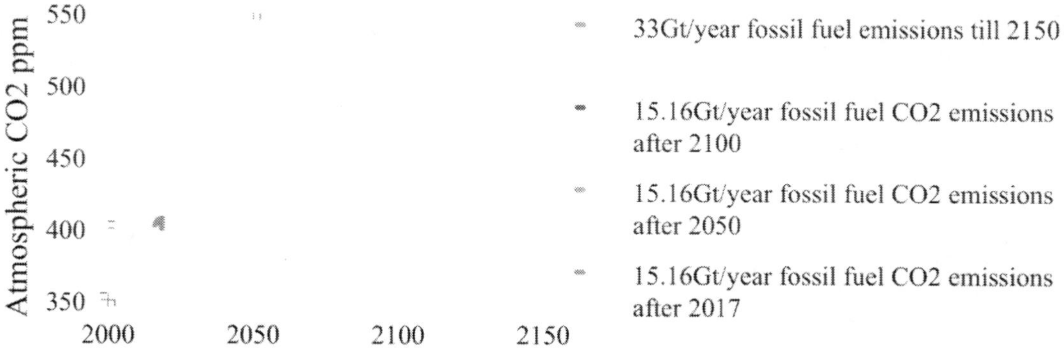

Figure 8 Atmospheric CO2 concentrations: 2000 to 2150

Prediction of Mean Annual Global Surface Temperatures from 2018 to 2100

The regression of five year averages of NASA GISS mean annual anomalies (9) against mean January atmospheric CO_2 concentrations are shown in figure 9. The trend line, derived for atmospheric CO_2 between 278ppm and 406ppm is extrapolated to 1,000ppm in figure 9a. The regression R^2 of 0.93 for the NASA data suggests that the linear equation is valid over the measured temperature and CO_2 ranges. For the extrapolation, it has been assumed the relationship will remain linear to 1,000ppm of atmospheric CO_2. This simple extrapolation is not meant to replace a detailed analysis of radiation absorption and emission by CO_2 in the atmosphere. However, its use results in the following estimates. The temperatures predicted in figure 10 for 2017 to 2100 are based on the atmospheric CO_2 concentrations predicted earlier in figure 7 of this paper. The 2000 to 2017 anomalies are from NASA GISS data. Results are similar to IPCC projections (10) made using similar CO_2 emissions' rates. If fossil fuel CO_2 emissions remain at 33Gt/year, close to their current level, the Earth's surface temperature will increase by just less than 1°K between 2018 and 2100. If CO_2 emissions are reduced to 15.2Gt/year there will be no increase in temperature. If fossil fuel CO_2 emissions rates increase at their pre 2011 average of 1.8%/year, the global mean surface temperature will increase 5°K by 2100. Completely eliminating fossil fuel emissions would bring the temperature anomaly increase back to 0°K, which is approximately a 1°K reduction from the current temperature.

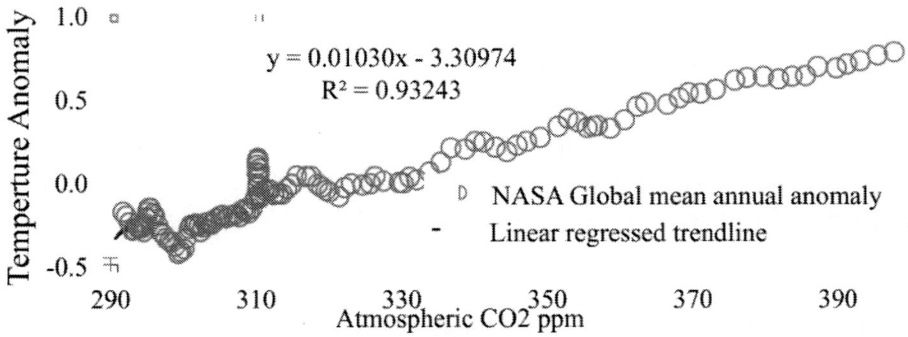

Figure 9 Mean Annual Global Temperature Anomalies

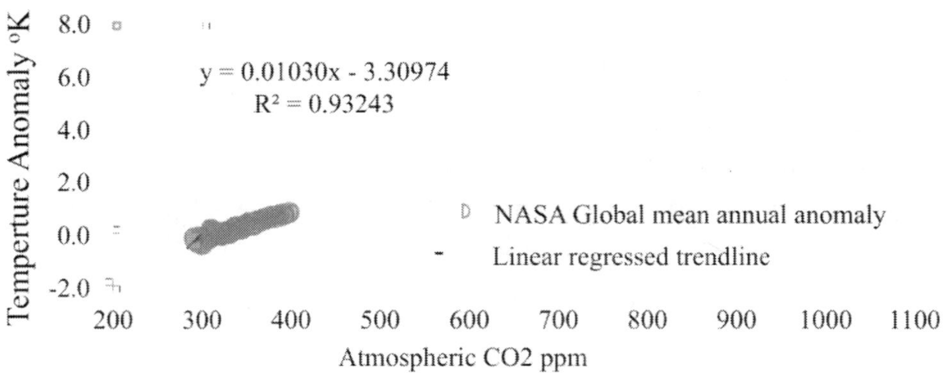

Figure 9a Mean Annual Global Temperature Anomalies

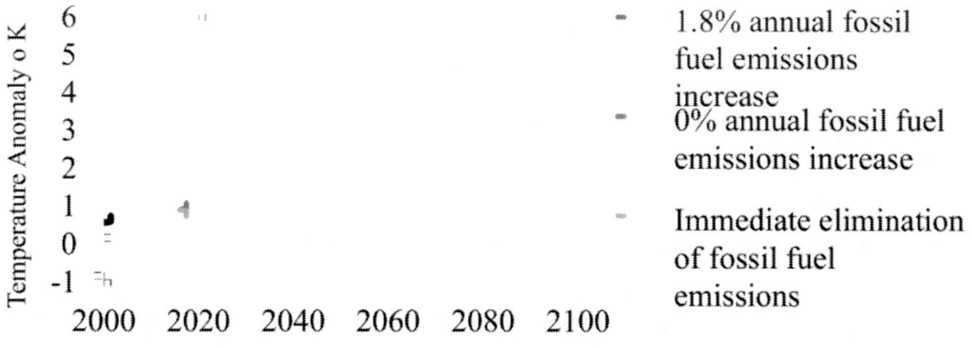

Figure 10 Temperature Anomaly: 2018 to 2100

Effects of Seawater Absorption on Atmospheric Carbon Dioxide Concentrations

The oceans are not in equilibrium with the atmosphere. It is this lack of equilibrium that facilitates CO_2 absorption. Without photosynthesis the oceans' surface waters would usually be in equilibrium and the CO_2 vapor pressure would equal the atmospheric CO_2 partial pressure. However, changes in temperature can result in the water being out of equilibrium, even without photosynthesis, but if this happens, CO_2 absorption or desorption will bring them back. Any diurnal or seasonal temperature effects are assumed to be insignificant and self-canceling over the calendar year time frame used for calculating average absorption rates and rate constants.

Photosynthesis removes CO_2 from the surface water, whereas absorption from the atmosphere, biomass respiration, decay and decomposition add CO_2. These mechanisms are functions of CO_2 concentration, temperature, sunlight availability and intensity, and the density of surface plants, plankton and nutrients. Some of the carbohydrates produced by photosynthesis are ultimately converted to calcium carbonate in shells, fish bones, and coral reefs. The global average CO_2 vapor pressure (**$VPCO_2$**) was approximately 93% of the atmospheric CO_2 partial pressure (CO_2) between 1990 and 2012 NOAA (11). The average measured difference between CO_2 and $VPCO_2$ (ΔCO_2) was 7% of the atmospheric CO_2 partial pressure. Therefore, $VPCO_2 = 0.07CO_2$. ΔCO_2 at the sample s listed in figure 12 was 28ppm. This was the average absorption driving force between the atmosphere and the oceans.
The absorption/desorption rates and $VPCO_2$ were determined experimentally using a sealed vessel containing the water sample with a pH meter, and a gas phase CO_2 meter. The vessel's atmospheric volume was 3.9×10^{-3} m^3, the water volume was 1.0×10^{-3} m^3 and water surface area was 3.6×10^{-2} m^2. All absorption and desorption tests were done at 25°C. PCO_2 denotes the CO_2 concentration in the gas phase of the sealed container. The initial PCO_2 in the vessel was between 50ppm and 4,000ppm

depending on the test. As time passed during each test, PCO_2 increased when the initial PCO_2 was lower than $VPCO_2$, and decreased when the initial PCO_2 was above $VPCO_2$. If there was no change in PCO_2, the initial PCO_2 equaled $VPCO_2$.This tedious method of determining $VPCO_2$ was later replaced with a simpler method, which calculated $VPCO_2$ from pH. The absorption/desorption rate was shown experimentally to be $r = k_{as}.a_s(PCO_2 - VPCO_2)$ for all samples. The seawater absorption rate constant for the sealed container k_{as} was calculated from measured values of r, PCO_2, $VPCO_2$, and PCO2. The seawater absorption surface area in the sealed container was a_s. The calculated rate constant was similar for samples from all locations. An absorption and desorption plot is shown in figure 11 for a seawater sample from the Gulf of Mexico at the Galveston San Luis Hotel breakwater. The measured rate constant $k_{as}.a_s$ was –0.14ppm/hour/ppm. The CO_2 in the sealed container could be calculated from $PCO_{2\,at\,time\,t} = e^{-0.14t}(PCO_{2c\,initial} - PCO_{2equilibrium}) + PCO_{2equilibrium}$ where "t" is time in hours of absorption or desorption. The calculated rate constant k_{as} for the sealed container was -6.77×10^{-7}mol/hr/m^2/ΔPCO_2 or -6.52×10^{-7} mt/year/m^2/ΔPCO_2. Assuming the rate constant in the sealed container is equal to the oceanic rate constant, and the oceanic $\Delta CO_2 = 0.07CO_2$, the oceanic rate constant k_{ao} becomes -4.57×10^{-8} mt/year/m^2/ppm of CO_2. Because 1ppm of atmosphere is 7.799Gt, this can be expressed as $k_{ao} = -5.86\times10^{-18}$ppm/year/m^2/ppm. Assuming the effective ocean absorption area is 8.21×10^{14}m^2, the oceanic rate constant inclusive of area $k_{ao}.a_o$ becomes –0.0047ppm/year/ppm. The effective ocean absorption area with waves and ripples was 8.21×10^{14}m^2 and not the waveless area of 3.67×10^{14}m^2. This assumes the oceans are covered with 2 feet high waves, each with a wavelength of 8 feet. In addition, each 2 feet high wave is assumed to be covered with 2 inch high ripples each with a wavelength of 8 inches. This assumption is based on observations made while taking the seawater samples.

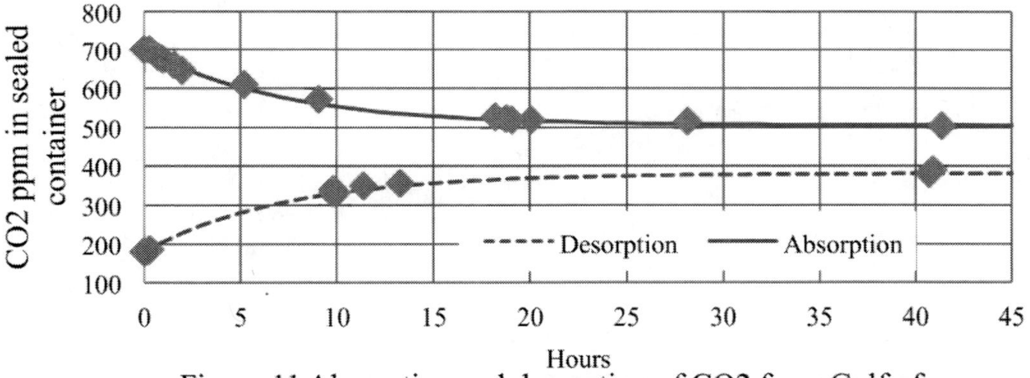

Figure 11 Absorption and desorption of CO2 from Gulf of Mexico water in sealed container

Ocean pH as a function of seawater CO_2 vapor pressures at 25°C

During seawater testing to measure CO_2 vapor pressure ($VPCO_2$) it was observed that $VPCO_2$ could be derived directly from seawater pH, saving many hours of testing per sample, and minimizing post sampling contamination. The pH vs. $VPCO_2$ is shown in figures 12, 13 and 14 for normal salinity (34-36g/kg) and low salinity (15-20g/kg). The Baltic and Black seas were the only low salinity samples. As pH is the log function of H^+ ion concentration, pH of seawater was a logarithmic function of $VPCO_2$. The average pH of normal salinity samples was $-0.42Ln(VPCO_2) - 10.55$ whereas the pH of the low salinity samples was $-0.39Ln(VPCO2) - 10.57$. The pHs of some recent samples from Miami Beach, Bahamas, Caribbean, Hawaii, and Antarctica all fit on the normal salinity curves. It is assumed that the low salinity samples have higher pHs because the ratio of Ca^{++}/Cl^- and Ca^{++}/Na^+ is 20% higher for the low salinity samples (12). $PCO_2 - VPCO_2$ or ΔCO_2 may be calculated from ΔpH using figure 15 for Galveston seawater. $\Delta pH = pH_e - pH_i$, where pH_i is the surface water pH measured in-situ and pH_e is the sample pH after it has equilibrated in the atmosphere at the sample site. If a CO_2 meter is not available, PCO_2 and the equilibrium $VPCO_2$ can be estimated from pH using figure 15. This is an exponential curve $\Delta CO_2 = 397 - e^{((\Delta pH - 2.5)/-0.42)}$. It may be approximated as linear over a $VPCO_2$ range of 300 to 500 ppm. The linear equation is $\Delta CO_2 = 766\,\Delta pH + 3.8$

Figures 12,13, 14 and 15 were derived at 25°C. Preliminary tests show that figures 12,13 and 14 accurately predict pH from vapor pressure between 20°C and 30°C. Figure 15 predicts pH accurately for temperatures as low as 0.4°C. To convert CO_2 vapor pressure to equivalent atmospheric CO_2 partial pressure, multiply the vapor pressure by 1.075.

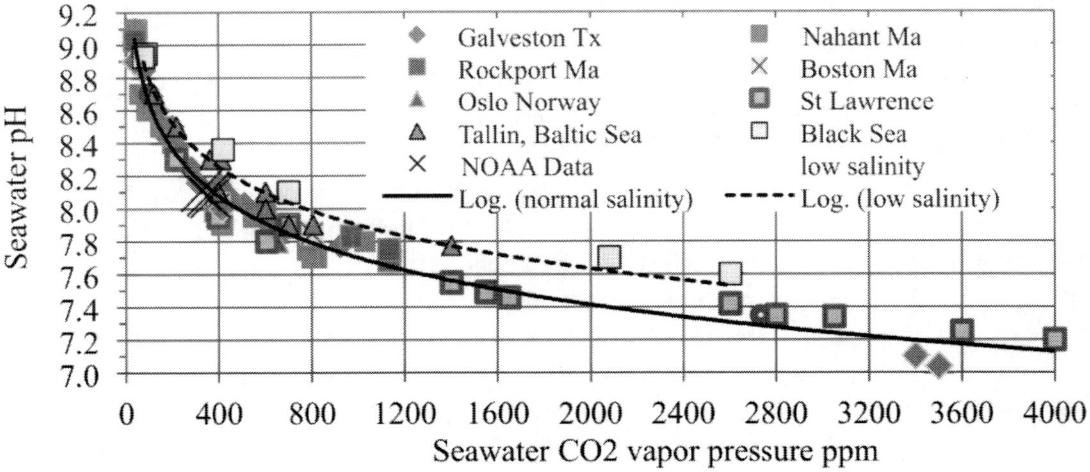

Figure 12 pH vs. CO2 vapor pressure

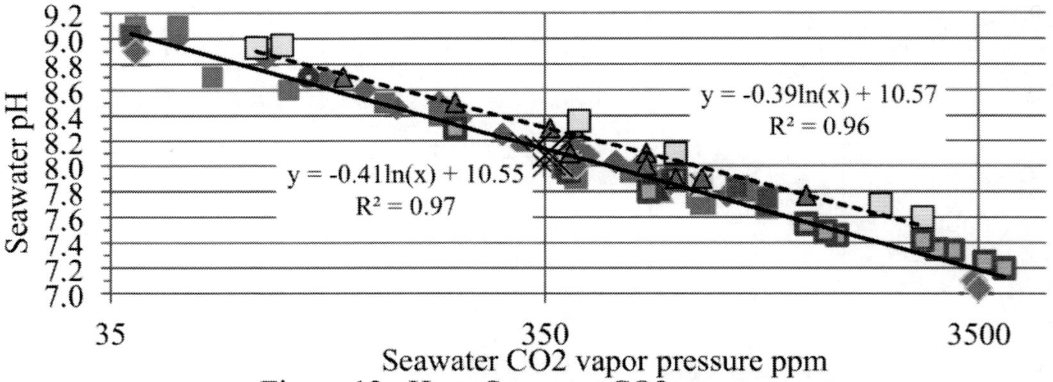

Figure 13 pH vs. Seawater CO2 vapor pressure

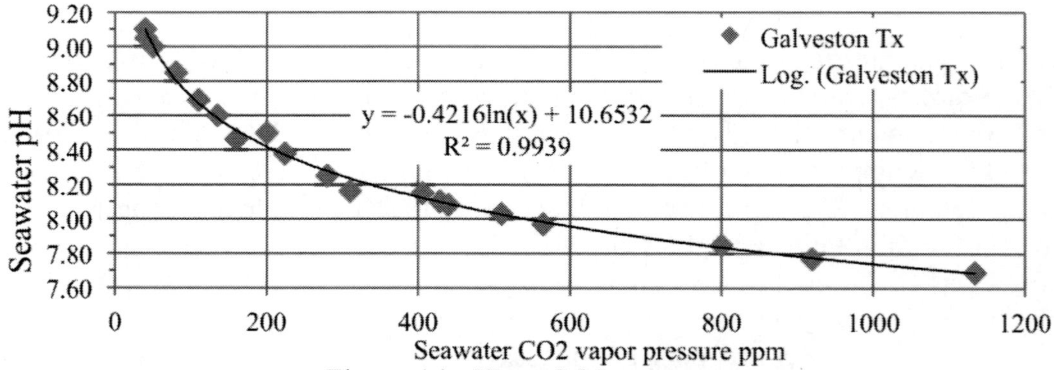

Figure 14 pH vs. CO2 vapor pressure

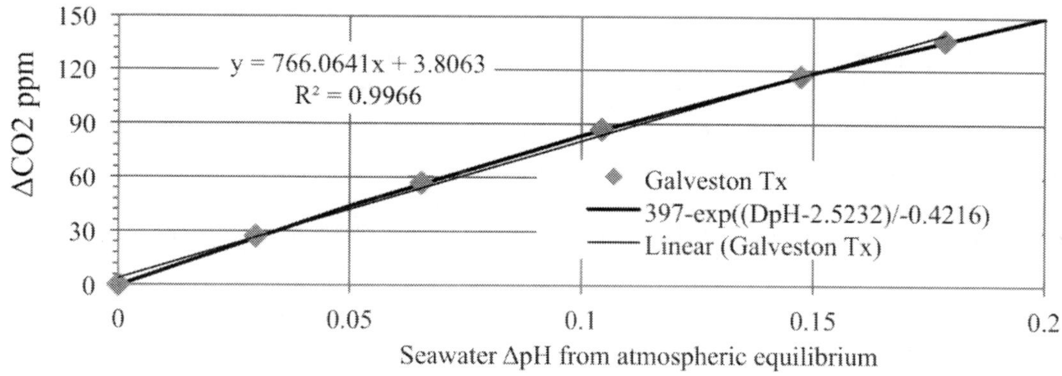

Figure 15 ΔCO2 vs. ΔpH

Estimated absorption rate constant for land-based biomass

Assuming the combined CO_2 removal rate from the atmosphere by land and oceans is $0.018PCO_2$ppm/year, the overall removal rate constant inclusive of absorption surface area is –0.018ppm/year/ppm. The overall net oceanic absorption rate constant $k_{ao}.a_o$ prorated from the measured seawater rate constant is –0.0047ppm/year/ppm. Therefore, the land-based rate constant inclusive of absorption area, $k_{al}.a_l = -0.0133$ppm/year/ppm. This suggests that land-based biomass absorbs at more than double the rate of the oceans. If the effective ocean absorption area a_s were different from the value used in this paper, the ratio of ocean to land-based rate constant would be different. However, this wouldn't affect the overall rate constant of –0.018ppm/year/ppm. The land-based rate constant k_{al} could not be estimated because a_l the land-based biomass absorption surface area was not measured or estimated.

Conclusions

Annual changes in atmospheric CO_2 concentrations are significantly effected by only two variables: fossil fuel emissions rate; and the atmospheric CO_2 concentration, The fossil fuel emissions rate is the only independent variable, because the atmospheric CO_2 concentration is a function of the fossil fuel emissions rate.

The equation $\delta CO_2 = \delta Ef + 5.13 - 0.018CO_2$ accurately represents global CO_2 mass transfer and is useful for evaluating the effects of CO_2 emissions on atmospheric CO_2 concentrations, surface temperatures and the pH of the oceans. The equation was derived from published data and supported by mass transfer theory and experimental results. δCO_2 is the annual change in atmospheric CO_2 concentration. δEf is the annual fossil fuel CO_2 emissions in ppm. The equation constant of 5.13 represents the annual ppm of non-fossil fuel CO_2 emissions, which include CO_2 from volcanoes, forest fires, and land-based biomass decomposition. $0.018CO_2$ represents the annual rate of CO_2 removal from the atmosphere by land-based biomass and the oceans. The net annual rate of CO_2 removal from the atmosphere is 1.8% of the total CO_2 in the atmosphere. The oceans absorb 0.5% of atmospheric CO_2 every year and 1.3% is absorbed by land-based biomass.

Atmospheric CO_2 concentrations correlate to global surface temperature anomalies.

Increases in atmospheric CO_2 cause decreases in oceanic pH. Low salinity seawater has a higher pH than high salinity seawater at all CO_2 vapor pressures due to its higher calcium ratios.

There were no significant changes in the CO_2 absorption rate constants, absorption surface area, photosynthesis surface area, photosynthesis rate constants, or non-fossil fuel CO_2 emissions between 1750 and 2016. Without the increase in fossil fuel CO_2 emissions atmospheric CO_2 would currently be

290ppm instead of 410ppm. If fossil fuel CO_2 emissions were reduced to their 1750 level, atmospheric CO_2 would decrease from 410ppm to 290ppm. This would take approximately 200 years because as atmospheric CO_2 decreases, the rate of CO_2 removal from the atmosphere also decreases.

If CO_2 emissions were reduced by 17.84Gt/year, atmospheric CO_2 concentration would equilibrate at the 2015 level of 400ppm. This may be achieved by reducing fossil fuel CO_2 emissions from 33Gt/year to 15.16Gt/year.

If future CO_2 absorption and photosynthesis surface area and kinetics remain similar to those from 1750 to 2017, and non-fossil fuel emissions remain constant, fossil fuel CO_2 emissions will be the only significant variable affecting future atmospheric CO_2 levels. The reduction of fossil fuel emissions is the only effective method of reducing atmospheric CO_2 concentrations. Only by reducing atmospheric CO_2 concentrations can the Earth's surface temperatures and the oceans' pHs be stabilized.

References

1. Octave Levenspiel, Chemical Reaction Engineering, Wiley & Sons 1964 p386
2. W. K. Lewis, W. G. Whitman Ind. Eng. Chem., 16, No. 12 1924
3. Fick, A., Ann. Phys. 94 (1855) 59. Diffusion
4. Carbon Dioxide Information Analysis Center (CDIAC) Historical CO_2 record from the Law Dome DE08, DE08-2, and DSS ice cores June 1998 D.M. Etheridge
5. National Oceanic and Atmospheric Administration (NOAA) Earth System Research Laboratory ESRL atmospheric CO_2 data published July 16 2012. Data added for 2012-2017
6. Energy Information Administration (EIA) International Reports 2006, 2009, 2011). International Energy outlook 201. 3 July 35 2013. Report Number: DOE/EIA-0484(2013)
7. Carbon Dioxide Information Analysis Center (CDIAC): Global CO_2 Emissions from Fossil-Fuel Burning, Cement Manufacture, and Gas Flaring: 1751-2010 July 30, 2013 Tom Boden, Carbon Dioxide Information Analysis Center Oak Ridge National Laboratory
8. International Energy Agency (IEA) May 24 2012 Newsroom
9. NASA GISTEMP Team, 2016: GISS Surface Temperature Analysis, Data set accessed 2018/01/30 at htpps://data.giss.nasa.gov/gistemp/.
10. Intergovernmental Panel on Climate Change. Fourth Assessment (AR4) 2007
11. NOAA's PMEL Carbon Program(www.pmel.noaa.gov/CO2)
12. Calcium Distribution in the world ocean waters, B. S. Krumgalz, OCEANOLOGICA ACTA 1982 Vol. 5 N° 1

512736 Operating Flexibility and Controllability of CO_2 Post-Combustion Capture Process with Different Amines

Abstract:

Carbon dioxide (CO_2) emissions capture from post-combustion is a key aspect in the stabilization of global climate change. Post-combustion CO_2 capture using amine-based solvents, typically monoethanolamine (MEA) and (Methyl diethanolamine) MDEA, is one of the most promising technologies for the CO2 emissions reduction. In practice, because power plants need to adjust their operation loads according to the peak and off-peak scenarios of power consumptions, the feed rate of flue gas from power plants features significant fluctuations. Thus, a CO_2 capture process has to consider its operating flexibility and controllability issues under disturbance of feed rate changes. In this study, a dynamic model of CO_2 capture process has been developed to study dynamic behaviors of the CO_2 capture process with different solvents (MEA or MDEA). Such a dynamic study is a must because MEA and MDEA have different CO_2 absorption capacities and reaction rates with CO_2, which will lead to different controllability of the process. In our study, the flue gas flow rate and reboiler heat duty are key process variables that determine the lean loading, liquid to gas ratio and percentage of CO_2 removal. The results show that the lean loading stream entering the absorber unit is critical to maintain the feasible operation of the system. Moreover, the energy performance of the process is directly affected by the operation of the reboiler. The outcomes from this study may provide more insights into the CO_2 capture process regarding the dynamic behavior of this process, which can be useful to help choose optimal amines for the specific process.

513527 Operating Condition Optimization of a Full-Scale Stirred Tank for Biofuel Production

Abstract:

Global warming and greenhouse gas emissions as well as the exhaustion of easily accessible fossil fuel resources are calling for effective carbon dioxide (CO2) mitigation technologies and clean and renewable energy sources. One promising technology is the conversion of CO2 by microalgae into biofuels. This process has several advantages but still suffers from economic feasibility.

In this study we are looking into this process by using Computational Fluid Dynamics (CFD) to understand and optimize the operating conditions in an aerated full-scale mixing vessel. The operating conditions are the key factors for a more efficient and economic process. The goal of the optimization is to reduce the power consumption for aeration and the impellers as well as to supply the best environment for the microalgae. Their growth and activity is affected by several parameters like the availability of dissolved CO2, sunlight (energy supply), further components, pH-value as well as shear forces, which can lead to significant cell damage.

Numerical pre-studies [1] have shown, that especially in full-scale the description of the interfacial area between the gas bubbles and the liquid phase is important for momentum, energy and mass transfer (CO2) and therefor for the amount of dissolved CO2. And this can typically only be achieved with a bubble size distribution including coalescence and breakup phenomena, which is considered in this study.

For the CFD simulation STAR-CCM+ by Siemens PL is used in combination with the design space exploration tool HEEDS to find the best configuration which matches the requirements. The requirements here are to minimize the power input but still to supply sufficient CO2 and avoid excessive shear acting on the microalgae. Based on the achieved results the Pareto-front is discussed and the effects on the efficiency of the process is shown. Furthermore the transferability to other reactor concepts is discussed.

[1] Aglave, R., & Eppinger, T. (2016, December). Influence of Bubble Size Distribution in Gas-Liquid Flows. In *High Performance Computing Workshops*. IEEE.

513596 Inorganic Salts Removal from Hydraulic Fracturing Flowback Water By Cooling

Abstract:

This paper reported a novel method for removing inorganic salts from hydraulic fracturing flowback water by cooling. Both model compounds and real samples were tested. Model compounds study shows that the cooling effect for the following five salts are in the order of

$KCl > Sr(NO_3)_2 > NaCl > CsCl > CaCl_2$.

The real samples show that about 3% of inorganic salts were removed by cooling. This method can be applied to hydraulic fracturing flowback water recycling. The method could be improved the separation of oil from the water before the test.

512841 Metal-Hexacyanoferrate Based Nanoadsorbents for the Removal of Radioactive Cesium from Seawater

Abstract:

Although the nuclear power has been proposed as an alternative to thermal power due to the concern over global warming, it is also perceived as a threat due to the potential for the uncontrolled release of radionuclides. For example, the nuclear accident at Japan in 2011 released a huge quantity of radioactive contaminants into the environment. Among these, 137Cs is the most problematic contaminant due to its long half-life (30.2 years) and high-energy gamma ray emissions. Although various types of adsorbents such as zeolites, crystalline silicotitanate have been already developed, these adsorbents are either expensive or inefficient for large-scale remediation due to their low Cs selectivity. Alternatively, transition-metal hexacyanoferrates including Prussian blue, have been effectively used to remove 137Cs after Chernobyl nuclear accident due to their low cost and high selectivity for Cs. In recent years, various Prussian blue-functionalized magnetic nanoparticles have been reported for the efficient removal of Cs and magnetic recovery of adsorbent after use. However, Cs removal in seawater remains a large challenge due to high salt concentrations, and since the sorption capacity of PB for Cs is less than the sorption capacities of other metal (such as Cu, Ni, Co, and Zn)-hexacynoferrates.
To overcome these shortcomings, in the present study, various metal hexacyanoferrate-functionalized nanoadsorbents having enhanced colloidal stability in seawater were fabricated for the treatment of contaminated seawater. The resulting a few nanoadsorbents exhibited excellent adsorption capacity for Cs and also showed excellent removal efficiency of radioactive cesium (137Cs) from contaminated seawater.

512885 A Computational Model to Simulate Performance of Industrial Clarifiers

Abstract:

Sedimentation is one of the most popular water treatment processes and is used to separate solid particles from carrier fluid in settling tanks known as clarifiers. The clarifier as the last major facility in waste water treatment plant (WWTP) can affect and sometimes even limit the overall performance of the WWTP. An industrial-scale clarifier investigated in this work experienced lower than designed performance, which limited its wastewater handling capacity. The interested clarifier has a non-conventional feedwell design. The objective of this study was to analyze the ramifications of the current clarifier design and to come up with design modifications to improve the performance. A novel 3-dimensional unsteady CFD (computational fluid dynamics) model, where an effective and simplified approach was employed to address solids removal and slow-rotating rake (at about 1 rotation per hour), was developed to analyze the clarifier performance. The root causes for the unsatisfactory performance of the current clarifier design were identified. Subsequently, various design modifications were developed using the model to significantly improve the clarifier performance. This work shows that the present numerical model can become a computationally effective tool that will help researchers to design and optimize complex industrial clarifier geometries with rotating rakes in order to reduce the effluent TSS (total suspended solids) and optimize the clarifier performance.

513245 CFD Modeling of Waste Water Bio-Reactors

Abstract:

Computational Fluid Dynamic (CFD) modeling is a powerful tool for equipment design and development and is widely used at UOP, particularly for the multiphase systems. One such system is UOPâ€™s XCeed™ Bioreactor technology. Through CFD modeling, the hydrodynamic interaction of the air and water in the fixed-bed bioreactor was studied and the results were used to design a new large-scale wastewater treatment bioreactor. An innovative and efficient modelling method was developed to overcome a complex bioreactor configuration.

The XCeed™ bioreactor packing media has a proprietary solid phase which uniquely distributes high surface area within highly porous void spaces.

The critical challenge in this three-phase CFD model was finding an accurate approach to mimic wastewater (hydraulic) and aeration (gas-bubbles) patterns through the solid phase containing substantial biomass.

To better understand the flow pattern, flow distribution and the air bubble residence time inside the XCeed™ Bioreactor, the analysis was broken down into two sub-models.

Part 1 single-phase water-flow only analysis showed that the XCeed™ packing media caused the water flow to distribute uniformly across the packed-bed, even when high levels of biomass are present, without any channeling or by-pass or plugging.

Part 2 two-phase air/water flow analysis showed that the XCeed packing void areas affected immediate air flow expansion into and subsequent distribution throughout the high surface area of the packed bed, even with the accumulation of high biomass.

These CFD results confirm the effective benefits of UOPâ€™s distinctive approach to blending surface area with open space to enhance air/water distribution without plugging or channeling even when high biomass solids are present.

513356 Optimization of Water Demand Growth Fulfillment in Seawater Thermal Desalination System

Abstract:

In arid regions of the world such as Saudi Arabia, thermal desalination technologies such as Multiple Effect Distillation (MED) are the mainstream technologies for producing desalinated water for both the residential and industrial sectors. Thermal desalination technologies are generally characterized by its high capital intensity. Fixed cost charges in MED typically account for 40-50% of the unit cost of production compared to 30% in RO [1]. This poses a challenge in distressed thermal desalination network with an incessant water demand growth as the conventional approach of large grassroots expansion projects can lead to low capital productivity.

With the emergence of new technologies such as Membrane Distillation (MD) that can integrated with existing conventional thermal desalination units to boost water recovery, the problem of water demand growth fulfillment in a desalination network extends beyond searching for the optimal capacity expansions to finding the optimal portfolio of projects in the planning horizon. The considered technologies vary in their capital intensity, modularity, and performance calling for an overarching optimization framework of the capacity-planning problem.

This work develops the optimization framework for the fulfillment of water demand growth in a distressed desalination network considering a set of technological alternatives which are both discrete and continuous in nature. A superstructure representation and optimization formulation are introduced to simultaneously optimize the investment portfolio in the desalination network for the minimization of the total net present value (NPV), which includes both annualized fixed and operating costs as well as salvage values, over a specified planning horizon. The framework is applied on a case study for the fulfillment of a linear water demand growth and an existing MED desalination system. The results are reported using a Pareto chart to show changes in the optimal projects portfolio with key parameters changes.

AUTHOR INDEX

AIChE
120 Wall Street, FL 23
New York, NY 10005-4020

ISBN 978-1-5108-6443-6

CURRAN ASSOCIATES INC.
proceedings.com